Automation

The McGraw-Hill *CONTROLLING PILOT ERROR* Series

Weather
Terry T. Lankford

Communications
Paul E. Illman

Automation
Vladimir Risukhin

Controlled Flight into Terrain (CFIT/CFTT)
Daryl R. Smith

Training and Instruction
David A. Frazier

Checklists and Compliance
Thomas P. Turner

Maintenance and Mechanics
Larry Reithmaier

Situational Awareness
Paul A. Craig

Fatigue
James C. Miller

Culture, Environment, and CRM
Tony Kern

Cover Photo Credits (clockwise from upper left): PhotoDisc; Corbis Images; from *Spin Management and Recovery* by Michael C. Love; PhotoDisc; PhotoDisc; PhotoDisc; image by Kelly Parr; © 2001 Mike Fizer, all rights reserved; *Piloting Basics Handbook* by Bjork, courtesy of McGraw-Hill; PhotoDisc.

CONTROLLING PILOT ERROR

Automation

Vladimir Risukhin

McGraw-Hill

New York Chicago San Francisco Lisbon London Madrid
Mexico City Milan New Delhi San Juan Seoul
Singapore Sydney Toronto

Library of Congress Cataloging-in-Publication Data

Risukhin, Vladimir.
  Automation / Vladimir Risukhin.
    p.   cm.
  ISBN 0-07-137320-9
  1. Airplanes—Automatic control. I. Title.

TL589.4.R57   2001
629.132'6—dc21   2001030503

McGraw-Hill

A Division of The McGraw·Hill Companies

Copyright © 2001 by The McGraw-Hill Companies, Inc. All rights reserved. Printed in the United States of America. Except as permitted under the United States Copyright Act of 1976, no part of this publication may be reproduced or distributed in any form or by any means, or stored in a data base or retrieval system, without the prior written permission of the publisher.

1 2 3 4 5 6 7 8 9 0   DOC/DOC   0 7 6 5 4 3 2 1

ISBN 0-07-137320-9

The sponsoring editor for this book was Shelley Ingram Carr, the editing supervisor was Stephen M. Smith, and the production supervisor was Pamela A. Pelton. It was set in Garamond following the TAB3A design by Michele Pridmore of McGraw-Hill's Hightstown, N.J., Professional Book Group composition unit.

Printed and bound by R. R. Donnelley & Sons Company.

McGraw-Hill books are available at special quantity discounts to use as premiums and sales promotions, or for use in corporate training programs. For more information, please write to the Director of Special Sales, Professional Publishing, McGraw-Hill, Two Penn Plaza, New York, NY 10121-2298. Or contact your local bookstore.

 This book is printed on recycled, acid-free paper containing a minimum of 50% recycled, de-inked fiber.

Information contained in this work has been obtained by The McGraw-Hill Companies, Inc. ("McGraw-Hill") from sources believed to be reliable. However, neither McGraw-Hill nor its authors guarantee the accuracy or completeness of any information published herein and neither McGraw-Hill nor its authors shall be responsible for any errors, omissions, or damages arising out of use of this information. This work is published with the understanding that McGraw-Hill and its authors are supplying information but are not attempting to render engineering or other professional services. If such services are required, the assistance of an appropriate professional should be sought.

*In memory of my father,
Nikolai Risukhin,
who gave me, together with my life,
the desire to do well everything I do*

Contents

Series Introduction *xi*

Foreword *xxi*

Preface *xxv*

Acknowledgments *xxxi*

Part 1 Acquaintance with Aircraft Automation

1 New Flight Operation Aspects Brought by Automation *3*
A Brief Introduction to Aircraft Automation *5*
Opportunities Brought by Automation *7*
Automation-Related Problems in Flight Operations *10*
Human-Factors Analysis in Modern Aviation Mishaps *14*
Pilot's Priority List: Automated Airplane Design and Operations Knowledge *33*

2 Automated Aircraft Design *35*
Automated Aircraft Controls *37*
Automatic Flight-Control System *42*
Automated Aircraft Power Plant *48*

Aircraft Motion Control Errors 55
Pilot's Priority List: Remaining in Continuous Control of the Aircraft Flight 62

Part 2 Automated Aircraft Crew Working Environment

3 Modern Flight Deck 67
Automated Aircraft Cockpit Technology 69
Aircraft System Status Electronic Indication 76
Isolation of Aircraft System Failures 85
Multifunctional Display 86
Aircraft-Status Monitoring Errors 89
Pilot's Priority List: Aircraft System Status Considerations 97

4 Flight Path Parameter Electronic Indications 99
Electronic Flight Instrument System 102
Primary Flight Display 103
Navigation Display 113
Electronic Flight Instrument System Control Panel 117
Crew Errors Resulting from Incorrect Use of Flight Parameter Indication 119
Pilot's Priority List: Procurement and Analysis of Flight-Relevant Information 129

5 Electronic Crew Warning Systems 131
Ground Proximity Warning System 134
Airborne Collision Avoidance System 142
Crew Errors in Use of Warning Systems 145
Pilot's Priority List: Immediate and Correct Reaction to Any Warning Signal 154

Part 3 Automated Aircraft in Flight

6 Flight Path Control 159
Manual Control of Aircraft Flight Path 162
Automatic Control of Aircraft Flight Path 163

Autopilot Flight Director and Autothrottle System
  Operation *163*
Automation Control Interface *173*
Errors in Flight Path Control *178*
Pilot's Priority List: Flight Path Monitoring and
  Control *183*

7 Automated Air Navigation *185*
Aircraft Navigation Electronic Systems *188*
Flight-Management-System-Controlled
  Navigation *194*
Navigation Errors *205*
Pilot's Priority List: Continuous and Complex
  Verification of Aircraft Position *221*

Part 4 Operating Aspects of Aircraft Automation

8 Human Role in Automated Aircraft Flight *225*
Task Allocation in Automated Flight *228*
Big-Team Collaboration *245*
Crew Coordination Errors *251*
Pilot's Priority List: Teamwork *257*

9 Securing Crew-Aircraft Automated System Efficiency *259*
Improvement of Pilot Literacy in Human
  Psychophysiology *263*
Securing Crew Proficiency *276*
Crew-Aircraft System Optimization *283*
Terms of Safe Automated Aircraft Flights *294*
Pilot's Priority List: Expanded Professionalism *297*
Summary *298*

Bibliography *301*

Index *309*

Series Introduction

The Human Condition

The Roman philosopher Cicero may have been the first to record the much-quoted phrase "to err is human." Since that time, for nearly 2000 years, the malady of human error has played out in triumph and tragedy. It has been the subject of countless doctoral dissertations, books, and, more recently, television documentaries such as "History's Greatest Military Blunders." Aviation is not exempt from this scrutiny, as evidenced by the excellent Learning Channel documentary "Blame the Pilot" or the NOVA special "Why Planes Crash," featuring John Nance. Indeed, error is so prevalent throughout history that our flaws have become associated with our very being, hence the phrase *the human condition*.

The Purpose of This Series

Simply stated, the purpose of the Controlling Pilot Error series is to address the so-called human condition, improve performance in aviation, and, in so doing, save a few lives. It is not our intent to rehash the work of over a millennia of expert and amateur opinions but rather to *apply* some of the more important and insightful theoretical perspectives to the life and death arena of manned flight. To the best of my knowledge, no effort of this magnitude has ever been attempted in aviation, or anywhere else for that matter. What follows is an extraordinary combination of why, what, and how to avoid and control error in aviation.

Because most pilots are practical people at heart— many of whom like to spin a yarn over a cold lager—we will apply this wisdom to the daily flight environment, using a case study approach. The vast majority of the case studies you will read are taken directly from aviators who have made mistakes (or have been victimized by the mistakes of others) and survived to tell about it. Further to their credit, they have reported these events via the anonymous Aviation Safety Reporting System (ASRS), an outstanding program that provides a wealth of extremely useful and *usable* data to those who seek to make the skies a safer place.

A Brief Word about the ASRS

The ASRS was established in 1975 under a Memorandum of Agreement between the Federal Aviation Administration (FAA) and the National Aeronautics and Space Administration (NASA). According to the official ASRS web site, *http://asrs.arc.nasa.gov*

> The ASRS collects, analyzes, and responds to voluntarily submitted aviation safety incident reports in order to lessen the likelihood of aviation accidents. ASRS data are used to:
> - Identify deficiencies and discrepancies in the National Aviation System (NAS) so that these can be remedied by appropriate authorities.
> - Support policy formulation and planning for, and improvements to, the NAS.

- Strengthen the foundation of aviation human factors safety research. This is particularly important since it is generally conceded *that over two-thirds of all aviation accidents and incidents have their roots in human performance errors* (emphasis added).

Certain types of analyses have already been done to the ASRS data to produce "data sets," or prepackaged groups of reports that have been screened "for the relevance to the topic description" (ASRS web site). These data sets serve as the foundation of our Controlling Pilot Error project. The data come *from* practitioners and are *for* practitioners.

The Great Debate

The title for this series was selected after much discussion and considerable debate. This is because many aviation professionals disagree about what should be done about the problem of pilot error. The debate is basically three sided. On one side are those who say we should seek any and all available means to *eliminate* human error from the cockpit. This effort takes on two forms. The first approach, backed by considerable capitalistic enthusiasm, is to automate human error out of the system. Literally billions of dollars are spent on so-called human-aiding technologies, high-tech systems such as the Ground Proximity Warning System (GPWS) and the Traffic Alert and Collision Avoidance System (TCAS). Although these systems have undoubtedly made the skies safer, some argue that they have made the pilot more complacent and dependent on the automation, creating an entirely new set of pilot errors. Already the automation enthusiasts are seeking robotic answers for this new challenge. Not surprisingly, many pilot trainers see the problem from a slightly different angle.

Another branch on the "eliminate error" side of the debate argues for higher training and education standards, more accountability, and better screening. This group (of which I count myself a member) argues that some industries (but not yet ours) simply don't make serious errors, or at least the errors are so infrequent that they are statistically nonexistent. This group asks, "How many errors should we allow those who handle nuclear weapons or highly dangerous viruses like ebola or anthrax?" The group cites research on high-reliability organizations (HROs) and believes that aviation needs to be molded into the HRO mentality. (For more on high-reliability organizations, see *Culture, Environment, and CRM* in this series.) As you might expect, many status quo aviators don't warm quickly to these ideas for more education, training, and accountability—and point to their excellent safety records to say such efforts are not needed. They recommend a different approach, one where no one is really at fault.

On the far opposite side of the debate lie those who argue for "blameless cultures" and "error-tolerant systems." This group agrees with Cicero that "to err is human" and advocates "error-management," a concept that prepares pilots to recognize and "trap" error before it can build upon itself into a mishap chain of events. The group feels that training should be focused on primarily error mitigation rather than (or, in some cases, in addition to) error prevention.

Falling somewhere between these two extremes are two less-radical but still opposing ideas. The first approach is designed to prevent a reoccurring error. It goes something like this: "Pilot X did this or that and it led to a mishap, so don't do what Pilot X did." Regulators are particularly fond of this approach, and they attempt to regulate the last mishap out of future existence. These so-called rules written in blood provide the traditionalist with plenty of training materials and even come with ready-made case studies—the mishap that precipitated the rule.

Opponents to this "last mishap" philosophy argue for a more positive approach, one where we educate and train *toward* a complete set of known and valid competencies (positive behaviors) instead of seeking to eliminate negative behaviors. This group argues that the professional airmanship potential of the vast majority of our aviators is seldom approached—let alone realized. This was the subject of an earlier McGraw-Hill release, *Redefining Airmanship*.[1]

Who's Right? Who's Wrong? Who Cares?

It's not about *who's* right, but rather *what's* right. Taking the philosophy that there is value in all sides of a debate, the Controlling Pilot Error series is the first truly comprehensive approach to pilot error. By taking a unique "before-during-after" approach and using modern-era case studies, 10 authors—each an expert in the subject at hand—methodically attack the problem of pilot error from several angles. First, they focus on error prevention by taking a case study and showing how preemptive education and training, applied to planning and execution, could have avoided the error entirely. Second, the authors apply error management principles to the case study to show how a mistake could have been (or was) mitigated after it was made. Finally, the case study participants are treated to a thorough "debrief," where alternatives are discussed to prevent a reoccurrence of the error. By analyzing the conditions before, during, and after each case study, we hope to combine the best of all areas of the error-prevention debate.

A Word on Authors and Format

Topics and authors for this series were carefully analyzed and hand-picked. As mentioned earlier, the topics were taken from preculled data sets and selected for their relevance by NASA-Ames scientists. The authors were chosen for their interest and expertise in the given topic area. Some are experienced authors and researchers, but, more importantly, *all* are highly experienced in the aviation field about which they are writing. In a word, they are practitioners and have "been there and done that" as it relates to their particular topic.

In many cases, the authors have chosen to expand on the ASRS reports with case studies from a variety of sources, including their own experience. Although Controlling Pilot Error is designed as a comprehensive series, the reader should not expect complete uniformity of format or analytical approach. Each author has brought his own unique style and strengths to bear on the problem at hand. For this reason, each volume in the series can be used as a stand-alone reference or as a part of a complete library of common pilot error materials.

Although there are nearly as many ways to view pilot error as there are to make them, all authors were familiarized with what I personally believe should be the industry standard for the analysis of human error in aviation. The Human Factors Analysis and Classification System (HFACS) builds upon the groundbreaking and seminal work of James Reason to identify and organize human error into distinct and extremely useful subcategories. Scott Shappell and Doug Wiegmann completed the picture of error and error resistance by identifying common fail points in organizations and individuals. The following overview of this outstanding guide[2] to understanding pilot error is adapted from a United States Navy mishap investigation presentation.

Simply writing off aviation mishaps to "aircrew error" is a simplistic, if not naive, approach to mishap causation. After all, it is well established that mishaps cannot be attributed to a single cause, or in most instances, even a single individual. Rather, accidents are the end result of a myriad of latent and active failures, only the last of which are the unsafe acts of the aircrew.

As described by Reason,[3] active failures are the actions or inactions of operators that are believed to cause the accident. Traditionally referred to as "pilot error," they are the last "unsafe acts" committed by aircrew, often with immediate and tragic consequences. For example, forgetting to lower the landing gear before touch down or hotdogging through a box canyon will yield relatively immediate, and potentially grave, consequences.

In contrast, latent failures are errors committed by individuals within the supervisory chain of command that effect the tragic sequence of events characteristic of an accident. For example, it is not difficult to understand how tasking aviators at the expense of quality crew rest can lead to fatigue and ultimately errors (active failures) in the cockpit. Viewed from this perspective then, the unsafe acts of aircrew are the end result of a long chain of causes whose roots originate in other parts (often the upper echelons) of the organization. The problem is that these latent failures may lie dormant or undetected for hours, days, weeks, or longer until one day they bite the unsuspecting aircrew....

What makes [Reason's] "Swiss Cheese" model particularly useful in any investigation of pilot error is that it forces investigators to address latent failures within the causal sequence of events as well. For instance, latent failures such as fatigue, complacency, illness, and the loss of situational awareness all effect performance but can be overlooked by investigators with even the best of intentions. These particular latent failures are described within the context of the "Swiss Cheese" model as preconditions for unsafe acts. Likewise, unsafe supervisory prac-

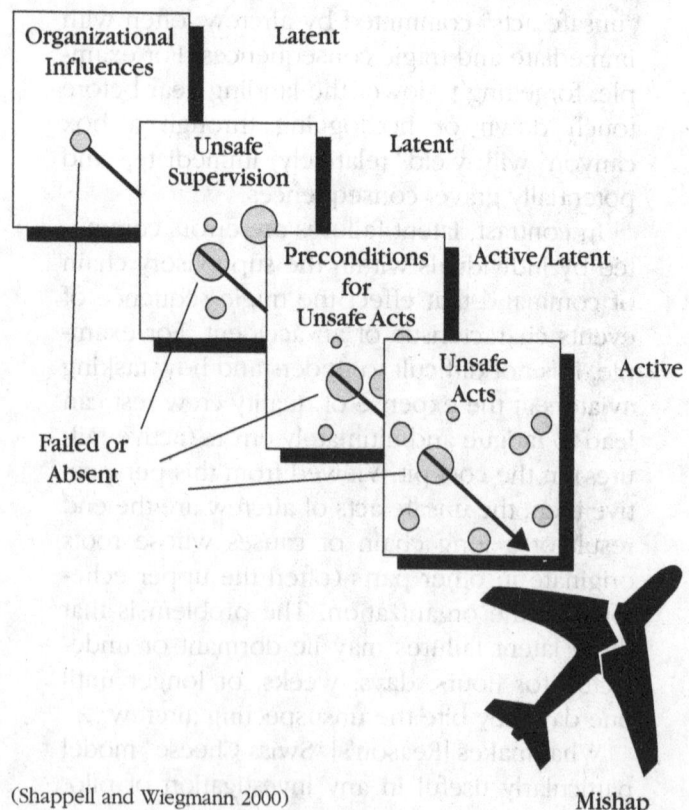

(Shappell and Wiegmann 2000)

tices can promote unsafe conditions within operators and ultimately unsafe acts will occur. Regardless, whenever a mishap does occur, the crew naturally bears a great deal of the responsibility and must be held accountable. However, in many instances, the latent failures at the supervisory level were equally, if not more, responsible for the mishap. In a sense, the crew was set up for failure....

But the "Swiss Cheese" model doesn't stop at the supervisory levels either; the organization itself can impact performance at all levels. For instance, in times of fiscal austerity funding is often cut, and as a result, training and flight time are curtailed. Supervisors are therefore left with tasking "non-proficient" aviators with sometimes-complex missions. Not surprisingly, causal factors such as task saturation and the loss of situational awareness will begin to appear and consequently performance in the cockpit will suffer. As such, causal factors at all levels must be addressed if any mishap investigation and prevention system is going to work.[4]

The HFACS serves as a reference for error interpretation throughout this series, and we gratefully acknowledge the works of Drs. Reason, Shappell, and Wiegmann in this effort.

No Time to Lose

So let us begin a journey together toward greater knowledge, improved awareness, and safer skies. Pick up any volume in this series and begin the process of self-analysis that is required for significant personal or organizational change. The complexity of the aviation environment demands a foundation of solid airmanship and a healthy, positive approach to combating pilot error. We believe this series will help you on this quest.

References

1. Kern, Tony, *Redefining Airmanship*, McGraw-Hill, New York, 1997.

2. Shappell, S. A., and Wiegmann, D. A., *The Human Factors Analysis and Classification System—HFACS*, DOT/FAA/AM-00/7, February 2000.

3. Reason, J. T., *Human Error*, Cambridge University Press, Cambridge, England, 1990.

4. U.S. Navy, *A Human Error Approach to Accident Investigation*, OPNAV 3750.6R, Appendix O, 2000.

Tony Kern

Foreword

The issues associated with automation in the cockpit are in their infancy. From the general aviation pilot working for the first time with a hand-held GPS to the experienced captain of a B777, change is the only constant. Anyone who has grown up in the so-called information age should clearly see that the pace of computerization and automation has accelerated to a nearly unmanageable rate in many sectors of industry. Transportation systems are not immune to this digital age danger, and for this reason it gives me great pleasure to write the Foreword for this outstanding contribution to the Controlling Pilot Error series.

The relationship between man and machine is wrought with complexity. To some, automation represents the first step in the elimination of human participation in the workplace. In many parts of the industrial world, this has already occurred. Consider the automated automobile assembly line or computerized mining operations that have displaced thousands of workers while simultaneously improving effectiveness and efficiency and, by the way, safety.

Some would argue that it is time to do the same in aviation. They cite the high percentage of human error mishaps as reason enough to move toward reducing the human contribution—if not the outright elimination of the pilot. We know that the technology already exists that would allow us to take the human being out of the cockpit. But that would be the wrong approach and is emphatically not the position of this book or myself. Let me explain why I feel this way.

The greatest supercomputer on earth cannot come close to the intuitive capacity of the educated, trained, and experienced pilot. When this pilot is integrated into an effective flight crew, the synergy increases its capacity and reliability, and when this team is fully integrated with an *appropriate level of human-aiding technology*, it should create a truly awesome capacity for safe and efficient operations. Yet this human-centered process does not always occur.

All too often, we read about automation that inhibits human performance. This occurs in a variety of ways, either through overreliance on automation that results in deterioration of pilot skills or perhaps by changing pilot workload, masking critical data, or just plain malfunction. The purpose of this book is to acquaint the reader with the fundamental issues of automation management and then provide a set of skills that will allow this understanding to be applied on the flight deck or in the cockpit.

As long as human beings remain in the cockpit, the pilot *must* be in control. The pilot must learn to use information and automation as an everyday part of his or her duties. He or she must learn to "listen" to automation much in the same way that we have taught modern flight crews to listen effectively to each other. To remain in control, the pilot must also know all of the capabilities of automation and use effective inquiry skills to stay situationally aware. The skilled pilot of tomorrow will use technology, naturally, but must always be prepared to exercise the judgment and the skill required to override the computer and take manual control of the situation when the need arises.

I have been honored to work with the author of this volume for several years. We have co-authored numerous articles on various airmanship topics, and I know him to be a warrior in the fight for safe skies. He possesses both theoretical and operational knowledge that only years of study and experience can provide.

Captain (Dr.) Risukhin knows the issues of automation as well as any operator in the world today. He has been actively involved in aviation instruction for over 25 years. More recently, he has specialized in taking pilots from large aircrews on traditional aircraft and training them to operate effectively in the modern two-person cockpits of Western aircraft such as the B767, B777, and Airbus. He is steeped in the art and science of human factors, and has conducted dozens of CRM courses with a focus on automation integration.

I now invite you to sit back and enjoy the wisdom and research of Russia's leading authority on aviation human factors and automation, and when you finish this book, pick up the rest of the Controlling Pilot Error series. I think it will make you a better pilot. Fly smart.

Tony Kern

Preface

Dear colleague!

The author calls you so because you have already introduced yourself by taking this book from the bookshelf. Only a person with an aviator's soul could be attracted by its name. Regardless of your present status in aviation, which may range from an aviation college student to professional instructor pilot to a happy airplane owner, you and the author are aviators, and he is extremely glad to meet you through this book. And as a true aviator, you are lucky. Why? Because the author has spent 10 years looking for the book, an aviator's concise manual on how to deal with automation.

In 1990 a leading Russian international airline by which the author was employed decided to start flight operations of automated aircraft. [Before that time, no pilot in Russia (the author's home country) could even think about flying a modern automated airliner.] Along with the introduction of the Russian-built automated aircraft Ilyushin 96 and Tupolev 204, preparation for the commercial utilization of the Western-built Airbus 310 was initiated.

The author was included in a group of the airline's instructor pilots selected for A310 transition training in France, Germany, and Canada, where aircraft of this type were built and operated. Before this training, we studied the English language intensively and tried to learn more about the principles and methods of automated aircraft flying. Therefore, we were eager for a book that would be accessible and easily understood yet would present sufficient technical detail to inform us as to the nature of aircraft automation, how automated aircraft are designed and controlled, what automatic systems are used in the cockpit, and how to operate them.

The book that could best answer those questions was the automated aircraft flight crew operations manual. It was big and heavy, contained a lot of data, and required sleepless nights for us to find and understand the needed information. Of course, aircraft designers and aviation authorities considered this to be the best book on automated aircraft available, but we were not aircraft designers and not even aviation top managers. We were rather experienced pilots who never dealt with automated aircraft and who had to soon become automated aircraft captains and instructors.

Even after becoming a captain of his first automated aircraft, the author continued to search for the book you are reading now. Although the instructors who had trained him in Toulouse, France, were the best in their profession, the training course was rather short, and many questions about the coordination between flight crew and aircraft automation emerged well after the training was over.

In 1994 the author was ordered to head a new group of Russian pilots to be trained in Seattle, Washington, for another automated aircraft, the Boeing 767. At this time the problem seemed to be less complicated. We had significant experience in flying previous automated aircraft and, we believed, knew the ideology of automated flying well. But our very first acquaintance with the new aircraft, in its flight simulator, showed us that another approach to the pilot-aircraft relationship could exist. The cockpit design and flight procedures proved to be simpler, but the flying abilities of the aircraft and crew were as high as in the previous automated aircraft operation. The flight operations manual at this time was much more concise. And again the author continued to search for a book in bookshops that contained answers to questions about common principles of design and organizing pilots' work in the automated cockpit.

In 1998, when the author began to fly and to teach other pilots to fly the most modern automated aircraft, the Boeing 777, the need for an easy-to-read yet professional-level book on aircraft automation design and flight operations basics once again intensified. During the past decade the author and other aviation professionals around the world were shocked by several automated aircraft accidents that occurred. It was known for sure that a lot had been done to properly train pilots of those aircraft, that the pilots were not unintelligent people, and that the aircraft were operable. But why had those deadly errors been committed?

In an attempt to answer this question, the author decided to arrange all the information he knew and could find about automated aircraft and their flight operations, and problems that had to be solved to fly those aircraft safely and efficiently. At that time, like you, dear reader, are now, the author was lucky. He obtained McGraw-Hill's proposal to write this book. Now you have it in your hands.

The book's contents are quite detailed because the use of automation in the aircraft cockpit has many aspects that must be described at least briefly to clarify potential problems. Although this book is short, it attempts to cover the most important issues of aircraft automation to help familiarize the reader with these issues in a systematic manner.

The text is divided into four logically connected parts. Part 1 (Chaps. 1 and 2) introduces automated aircraft to the reader. It shows how flight operations and aircraft design in general have changed with developments in aviation automation. Part 2 (Chaps. 3 to 5) is devoted to modern aircraft cockpit design and methods of flight-relevant information display for the pilot. Practical applications of aircraft automation during flight are described in Part 3 (Chaps. 6 and 7). Part 4 (Chaps. 8 and 9) tells the reader about crew activity during automated flight operations, and suggests possible ways of improving flight safety.

Chapter 1 shows the reader new opportunities and also new problems introduced by aircraft automation. A tool for aviation mishap analysis is also introduced.

Automated aircraft structure and its main components are described in Chap. 2.

Chapter 3 opens the major part of the book. To safely operate any airplane, the pilot must know the airplane cockpit well. Automated aircraft cockpit development and aircraft systems operation indication means are presented in that chapter.

Flight path parameter electronic indications (displays) are described in Chap. 4. Two main glass cockpit devices—the primary flight display and the navigation display—are the main topics.

Timely warnings to flight crews about flight safety problems have proved to be vitally important in contemporary aviation. Electronic crew warning systems and their operation during flight are described in Chap. 5.

In Chap. 6 the reader will learn in a systematic way how the automated aircraft flight path is controlled.

Possible ways of flight path control together with cockpit flight path control devices are described.

In Chap. 7 the reader will learn about automated aircraft navigation equipment and its operation. Practical applications of the flight management system and recommendations for automated aircraft navigation are also presented.

Chapter 8 tells the reader about automated flight task allocation, defining an optimal degree of automation utilization, and coordination between crew members. The roles of crew and other aviators in flight operations and how these functions can be improved in automated aircraft flight are described.

The final chapter of the book, Chap. 9, is devoted to securing automated flight efficiency. Specific physiological and psychological factors connected with the flight profession, crew training, and optimal design of the crew-aircraft system for improvements in flight safety are discussed. A summary of the future prospects of automated flight operations development concludes Chap. 9 and the book.

Flight crew errors connected with specific sides of automated aircraft flight operations are discussed, and relevant examples of aviation mishaps with automated aircraft, followed by analysis, are included in every chapter. The characteristics of each example correspond to the material described in the general text of the chapter. A short flight safety recommendation for the automated aircraft pilot concludes each chapter.

Professional terms and acronyms, which are inevitable in a book of this kind, are explained throughout.

Vladimir Risukhin

Acknowledgments

Among the people who gave me guidance for this book are instructors who made me familiar with automated aircraft: Peter Henny, Paul Dardin, and Jonathan Hutchinson of Airbus, and Stan Strong, Lonny White, Donald Smith, Jim Gill, William Roberson, and Robert Rugarber of Boeing. I highly appreciate their desire to give me the best they knew and could, and I am grateful to all of them for their professional teaching and training.

For years I was lucky to work side-by-side with leading Russian pilots who did a lot to introduce automated aircraft flight operations in my native country. The enthusiasm and professionalism of these people inspired my desire to write this book. Among them are Gennady Prikhodchenko, Dmitry Barilov, Victor Perepelitsa, Anatoly Volymerets, Vladimir Barilov, Vyacheslav Jelinsky, Stanislav Tulsky, Nikolai Puzyrev, Victor Sajenin, Anatoly Yakimchuk, and Bladimir Orlovsky. I thank all of them for their support and friendship.

Dr. Tony Kern has shown me a new direction where the true aviator should apply the professional abilities acquired during a long pilot's career. Thank you very much, dear Tony.

Ms. Shelley Ingram Carr, my sponsoring editor, organized my work with the book and gave me useful instructions and advice. Thank you very much for that, Ms. Carr.

Mr. Stephen M. Smith, my editing supervisor, provided me with valuable assistance during the book production process. I highly appreciate your help, Mr. Smith.

My mother, Nina Risukhina, gave me life and loves me always. Her love gave me the energy to do my best in writing this book. Thank you very much, mama.

My family has supported all my actions during the last 31 years. Together we have lived through every chapter of this book. My dear wife Tanya, son Dmitry, and daughter Katya, I love you and thank you very much for your love, interest, and assistance.

Part 1
Acquaintance with Aircraft Automation

1

New Flight Operation Aspects Brought by Automation

1

New Flight Operation Aspects Brought by Automation

Trends in world economic growth and free-market competition determine an airline's need to increase its profitability. Profitability can be increased by modification of two factors connected with aircraft operations: reducing direct operating costs, such as fuel consumption and maintenance; and increasing revenue potential defined in terms of operational capability, customer appeal, and infrastructure flexibility. The introduction of new technologies is the most appropriate tool with which to pursue these goals.

A Brief Introduction to Aircraft Automation

The industrial revolution significantly impacted human society, particularly by reducing the need for human manual labor. In a similar way, the computer revolution is now impacting human ability to process information and make decisions. Aviation, as one of the most technology-sensitive industries, is now significantly affected by these changes.

6 Acquaintance with Aircraft Automation

Around the late 1930s, a gyroscopic flight-control device called the *autopilot* (or *automatic pilot*) was first applied in aviation. By accurately controlling the airplane's steering and directional attitude, this automatic device enabled the human pilot to devote more attention to other important duties. This revolutionary step in flight-control technology marked the beginning of a new period in aviation history: the era of flight automation. The term *automation* is defined as "the use of automatic equipment and machines to do work previously done by people" (Hornby 1995). The term *automatic* in this context means "working by itself without direct human control" (Hornby 1995).

Introduction of the jet engine has significantly enhanced aircraft performance. This development, combined with the substantial growth in air-traffic density, has led to requirements for a higher rate of information processing within the flight crew cockpit. Strict flight path limitations imposed by increased commercial air traffic required more reliable communication between flight crews, autopilots, and air navigation radio facilities.

The progressive introduction of automatic devices capable of performing the functions of control, computation, and communication has demonstrated the potential of aircraft automation to provide greater safety and economy.

The last three decades of the twentieth century witnessed rapid progress in aircraft automation, which brought dramatic changes to the aviation industry. These days, for example, a large commercial aircraft typically is operated by only two human pilots and carries several hundred people or a dozen or so tons of cargo between continents without any intermediate landing or fuel stop. Another amazing development has become routine: aircraft automatic final runway approach, landing, and rollout. Without automation, it would be unthinkable to provide the current precision of aircraft navigation and flight path monitoring, which meets the requirements imposed by air-traffic density and significantly reduced weather landing minima.

Nevertheless, there is some controversy regarding the value of automation in the aviation industry. Proponents state that insufficient automation would impede further development of aviation technology and might invite human error in the industry. Opponents of this position insist that modern aircraft automation is unreliable and only complicates human operator performance in complex aviation systems.

This longstanding argument is illustrated by the implementation history of the *ground proximity warning system* (GPWS) (Braune and Fadden 1987). The use of this system in flight decks was made mandatory in 1974 to avoid multiple aircraft-terrain collisions. Although pilots frequently complained about false alarms highlighting automation failures, this system proved to be extremely useful. There is no documented evidence of how many potential accidents have been prevented because of the presence of the GPWS. However, statistics show that the GPWS and other automatic devices have made a significant positive contribution to flight safety.

In general, introduction of automatic equipment in aviation brought opportunities and problems that were unknown before.

Opportunities Brought by Automation

The technological progress realized with aircraft automation has had a positive impact on the flight accident rate. Each generation of aircraft has brought with it a significant improvement in flight safety. This phenomenon was initially observed after wood and fabric were replaced by steel and aluminum in the aircraft interior. Later, the emergence of turboprop and jet engines led to an even more dramatic increase in aircraft reliability. These days we can see that the flight safety record of modern highly automated aircraft is much better than that recorded for aircraft in the past.

Aircraft automation has made adherence to flight routes and schedules more economically feasible by improving air navigation precision and reducing flight operations weather minima. It has allowed crews to better comply with speed and altitude restrictions and has expanded aircraft-ground communication capabilities.

From the very beginning of the expansion of automation in aviation, a series of studies (Braune and Fadden 1987) confirmed the usefulness of automation in flight operations. For example, a study of flight crews operating a highly automated Boeing 757 aircraft proved that pilots responded very positively to the airplane and its automation; this response was reflected in their comments. Another study of advanced technology transport aircraft, conducted by the Aviation Safety Reporting System (ASRS), a cooperative program funded by the Federal Aviation Administration (FAA) and administered by the National Aeronautics and Space Administration (NASA), has shown some interesting figures in favor of automation. Pilots were asked to rate, on a scale of 1 to 10, several automated flight-deck features directly influencing flight performance, the automation on their airplane as a whole, and the airplane's independence from any constraints imposed by the air-traffic environment. For 13 flight-deck systems, including the flight management system (FMS), the autothrottle system, the electronic flight instrument system (EFIS), and their panel displays, an average score of 8.1 was obtained. The average score for the total aircraft automatics was 7.82. For the automated aircraft as an independent unit of the air-traffic environment, the average rating was 8.56.

Why do pilots who fly automated aircraft love them? It is not just the pleasure of using new technology, although this is a factor. A prime reason lies in EFIS's ability to display exactly what pilots want and when they want it. On an instrumental approach to land, for example, the EFIS attitude indicator shows glideslope and localizer scales, decision height setting and warning, actual airspeed and the speed-trend vector (indicating changes in airspeed), and other data needed to perform a successful landing. When the aircraft is in cruise mode, most of this information vanishes, leaving just the speed indications, with director bars added to roll and pitch scales showing a horizon symbol between the sky and earth images. The FMS produces all needed calculations, suggests a variety of flight management options, and maintains the flight path chosen by the pilot.

The potential of aircraft automation is amazing. The computer industry has shown that almost anything can be portrayed on instrumentation panel displays, and as aircraft automation enters more widespread use, the ability to introduce more novel presentations becomes a reality. *Speed-trend vectors,* which represent one of the earliest useful categories of additionally displayed information, immediately alert pilots to changing flight situations and prompt them to take early corrective action. These days pilots are accustomed to seeing cockpit panel displays indicating speed limitations, wind direction (including windshear potential) and velocity, weather formation, other traffic in the vicinity (including information reflecting collision threat potential), electronic approach charts, the electronic checklist, and even the terrain they are flying over. The computer memory contains details of the terrain over many thousands of square miles. Using automatic inputs of position, altitude, and heading, the computer creates a correct perspective image of the view ahead. The interpretation is instinctive and natural, handy for low-altitude and poor-visibility operations.

In general, automation has made the modern aviation picture bright, colorful, and very attractive. The flight in a comfortable Boeing 777 airliner is a real pleasure not only for most passengers, but also for most pilots operating this machine—as one of those pilots, the author knows that for sure.

Modern commercial aviation is generally much safer than any other mode of transportation. But sometimes society is puzzled and even frightened by accidents of highly automated aircraft. Why do these accidents happen? What must be done to avoid aviation disasters? To come closer to answers to these questions, one should look more carefully at some new problems induced by automation.

Automation-Related Problems in Flight Operations

Although the future of automated aircraft seems very promising, limited only by the imagination of aircraft systems designers, the new technology has brought a new class of aviation mishaps: automation-related incidents and accidents. Flight safety data for advanced-technology aircraft are generally better than those for conventional transport aircraft. However, the previously rapid improvement in flight safety has slowly become scarcely noticeable. Numerous research studies show that the accident rate has leveled off since 1990. A significant increase in the number of accidents is forecasted as air traffic continues to grow, unless something is

New Flight Operation Aspects Brought by Automation

done to reverse this trend. This book is one attempt to improve understanding of—and present general guidelines for—automated aircraft flight safety.

The first highly automated aircraft, such as the A300-600, A310, Boeing 757, and Boeing 767, were designed and built in the late 1970s and early 1980s. The people who created those aircraft were true automation pathfinders in aviation. They did and continue to do everything possible to make aircraft more safe, efficient, and comfortable. Many aviation electronic technology improvements have been made since that time. Aircraft fleets continue to be provided with new equipment, such as the enhanced ground proximity warning system (EGPWS), the airborne collision avoidance system (ACAS), and satellite navigation. These systems have a significant potential for safety enhancement. Nevertheless, it is doubtful that the technology alone will suffice to keep the accident rate moving downward, or even level. It is a great pity to acknowledge that against the background of high automation redundancy and reliability, approximately three out of every four modern aircraft disasters occur because of less-than-satisfactory human performance.

There is evidence that automation does not necessarily reduce pilot workload. Sometimes it can even increase a pilot's attention to and involvement with monitoring and programming. Automation also changes the operator's role in human-machine systems. The human operator no longer immediately controls the machine but controls the automation. This change has proved to be very important. In aviation, it means that a person who has been trained as a pilot must continue to function as a pilot and, in addition, acquire new professional qualities that are sometimes very different from the ability to fly the airplane.

The frequency of automation-related aircraft incidents has prompted specialists in the aviation industry to define several typical pilot-error categories that caused the incidents (Sheehan 1995): data entry, indication monitoring, system workarounds overriding computer programs, and automation mode misapplication. Combined with other negative contributing factors such as pilot inattention, fatigue, or distraction, these errors may threaten flight safety.

Aviation safety statistics provide detailed information about pilot responses to aircraft automation. Although the responses generally favor automation, pilots sometimes point out definite problems with automatic systems. The most common problem areas are crew training, flight procedures, and automatic systems design, all of which can be traced in incidents and accidents of automated aircraft. The most alarming situations in flight operations occur when flight crews do not understand what the automation is doing and how to correct its malfunctions. The following case illustrates this point.

Case 1: Automated aircraft deadly disobedience in Nagoya, Japan

On April 26, 1994, an A300-600 aircraft performed a scheduled flight from Taipei, Taiwan, to Nagoya, Japan. While making an instrumental approach to runway 34 of Nagoya airport, under manual control, the aircraft's first officer inadvertently touched a small control lever located on the thrust lever's inner surface. This control, called the *go lever*, activates aircraft takeoff or go-around maneuvers.

This action changed a flight path computer mode of operation from approach to a go-around mode, and caused a sudden increase in engine thrust. As a result, the aircraft deviated above its normal glide path.

New Flight Operation Aspects Brought by Automation   *13*

To prevent an increase in aircraft pitch, in accordance with the captain's instructions, the first officer started to push the control wheel. In an attempt to restore the normal flight, both autopilots were subsequently engaged by the crew, with the go-around mode still operative. As a result, in accordance with aircraft automation logics, the horizontal stabilizer trim moved to its full nose-up position, causing an abnormal out-of-trim situation and resulting in a tendency for the airplane to fly nose up.

After this the crew initially tried to continue the landing approach. Then the captain judged that landing would be difficult, took the controls, and opted for the go-around mode. The aircraft began to climb steeply with a high-pitch-angle attitude.

On an aircraft of this type, when the autopilot is engaged in either landing or go-around mode, the pilot pushing or pulling the control wheel can override movement of the elevators performed by the autopilot. In this case, however, the autopilot automatic trim orders are not canceled, and the autopilot will move the horizontal stabilizer trim so as to maintain the aircraft on the scheduled flight path. If the pilot continues to override the landing or go-around mode, the aircraft will eventually reach an imbalanced out-of-trim condition. With regard to this hazardous situation, a special caution is provided in the flight crew operations manual (FCOM).

The flight crew did not understand the aircraft automation logic and did not carry out an effective recovery operation. The aircraft finally stalled, crashed, ignited, and was destroyed. On board were 271 persons; 256 passengers (including 2 infants) and 15 crew members. There were 264 fatalities; only seven seriously injured passengers survived.

In this flight the crew was unable to cope with a difficult situation caused by a combination of factors related to aircraft design as well as flight crew proficiency.

Human-Factors Analysis in Modern Aviation Mishaps

During the transition to new technology, a number of safety problems arose in the aviation industry. Pilots, aeronautical engineers, scientists, and aviation safety experts have expressed concerns about the safety of flight-deck automation. Dozens of books and hundreds of magazine articles about flight automation peculiarities have been published. There have been many aviation accident investigations, including analysis and causation schemes, but there has been little practical application of the findings from these studies to improve flight safety standards. The research findings that have actually led to flight safety improvements in modern aviation are the most valuable.

Analysis tools oriented toward practical results

Two approaches seem to be most appropriate for use in automated aircraft flight safety assessment and flight disaster prevention: a human-factors analysis and classification system created by Scott A. Shappell and Douglas A. Wiegmann, and an airmanship model developed by Tony Kern.

Shappell and Wiegmann (2000) developed one of the latest tools to enable one to answer the question "Why do aircraft crash?" In their human-factors analysis and classification system (HFACS), they consider human error to be only a starting point in examination of the underlying causes of flight accidents. They state that accidents cannot be attributed to a single cause or to a single individual. It is also difficult to identify a primary cause of an accident. As a rule, accidents in aviation are final results of several causes. Unsafe crew acts are often only the last links in this chain of erroneous events.

Kern (1997) has created a realistic and practical model of airmanship that ensures a level of flight crew situational awareness sufficient for efficient decision making.

For construction of their HFACS model, Shappell and Wiegmann have used the "Swiss cheese" model of human error, developed by James Reason to increase the performance reliability of power-plant personnel. This model consists of four levels of human failure, each influencing the next:

1. Unsafe acts of operators, which have ultimately led to the accident
2. Preconditions for unsafe acts, involving elements relevant to crew resource management (CRM): mental fatigue, poor communication, and poor coordination
3. Unsafe supervision, when an inadequate crew is scheduled for the flight
4. Organizational influences, including circumstances such as fiscal austerity or budget constraints, resulting in insufficient training, increased flight time, and reduced rest periods

Each failure level can be described as a slice of cheese, with its specific deficiencies forming holes in the slice. It can be assumed that all slices combine to create four levels of anti-accident defense, and that the coincidence of the holes in these four slices leads to an accident.

For example, a training program may be cut as a result of an airline's financial difficulties. Without good crew resource management training, the airline's crews have poor communications and coordination skills. Less-than-proficient pilots are scheduled to perform complex tasks. In this case an accident is extremely probable if, for instance, the crew encounters even minor aircraft equipment failure when making a landing approach in weather minima conditions with a low cloud base and a strong crosswind. Or, as shown above in case 1, an accident can occur if the crew members encounter—even under absolutely normal flight conditions (no turbulence, etc.)—peculiarity of aircraft automation that may be generally known within the aviation community but unfamiliar to them.

James Reason's Swiss cheese model revolutionized common views of accident causation but did not clearly describe the implication of the "holes." To ensure flight safety improvement, the system failures have to be identified, detected, and corrected before the accident occurs.

In their HFACS work, Shappell and Wiegmann have defined the contents of the holes of Reason's model using a large amount of information about real aviation accidents. They described in detail the four failure levels and classification factors inherent to each level. Unsafe acts are at the first level. They can be classified into two categories: errors and violations.

Errors are identified as results of human mental or physical activities that fail to achieve their intended outcome. Errors dominate most aviation accident databases because they are caused by human beings, who are (theoretically) inherently error-prone. *Violations* are connected with willful disregard for the rules and regulations that govern flight safety. The categories of errors and violations are further expanded. Errors are divided into three basic types: decision, skill-based, and perceptual. Violations include routine and exceptional forms.

New Flight Operation Aspects Brought by Automation 17

Decision errors represent intended actions that prove inadequate or inappropriate for the given situation. These actions, called "honest mistakes," are the unsafe results of a pilot's insufficient knowledge or poor judgment. Decision errors are the most heavily investigated of all types of error. They can be grouped into three general categories: procedural errors, poor choices, and problem-solving errors. *Procedural errors* occur when a situation is either not recognized or misdiagnosed, particularly when pilots are faced with highly time-critical emergencies. *Poor-choice errors* occur when there is no standard procedure corresponding to a particular situation. *Problem-solving errors* usually occur in situations where pilots must quickly solve problems that they do not clearly understand.

Skill based errors in aviation are described as errors related to basic flight skills. They occur as a result of attention, memory, or flight technique failures. *Attention failures* are linked to inadequate visual scan patterns, task fixation, inadvertent activation of flight controls, and incorrect steps in sequencing a procedure. *Memory failures* appear as omitted checklist items, place losing, and forgotten intentions. The sources of *flight technique failures* can be traced to the manner in which pilots maneuver their aircraft. Failures of this type express pilots' personalities and may be difficult to prevent and mitigate.

Perceptual errors are results of pilots' incorrect sensory inputs. They occur in cases of visual illusions or spatial disorientation. Some pilots, although trained to overcome these phenomena, fail to monitor their flight instruments and become deceived by incorrectly perceived feelings.

Violations, in contrast to errors, are a willful disregard for the rules and regulations that govern safe flight.

They occur less frequently because of the relatively high level of discipline in aviation and exist in two distinct forms: routine and exceptional violations. *Routine violations* are habitual and often tolerated by governing authorities. To eliminate this kind of violation, it becomes necessary to look further up the authority chain of command or hierarchy to identify persons who are not enforcing the rules. Exceptional violations appear as isolated departures from authority. They are particularly difficult to predict and deal with.

The second failure level in the HFACS scheme includes unsafe crew conditions that can potentially cause unsafe crew acts. These conditions, called *preconditions for unsafe acts*, consist of two major subdivisions: substandard conditions of operators and substandard practices of operators.

Substandard conditions of operators are divided into three categories: adverse mental states, adverse physiological states, and physical or mental limitations. *Adverse mental states* are very critical in aviation and must be accounted for in the causal chain of events that results in human error. The most important of these states are the loss of situational awareness, task fixation, distraction, and mental fatigue. Personality traits and pernicious attitudes such as overconfidence, ungrounded complacency, and misplaced motivation are also considered important. *Adverse physiological states* include impaired physiological state, medical illness, physiological incapacitation, and physical fatigue. These conditions can lead to visual illusions, spatial disorientation, or poor decision making. *Physical or mental limitations* refer to those instances when mission requirements exceed pilot capabilities. This category includes basic sensory and information processing abilities together with qualities needed in aviation, such as physical strength and stamina and the ability to make decisions quickly and respond effectively in life-threatening situations.

According to the HFACS authors (Shappell and Wiegmann 2000), *substandard practices of operators* include two categories: crew resource mismanagement and personal readiness. The *crew resource mismanagement* category was created to account for poor communication and coordination within the flight crew as well as between the crew and other aviation personnel (air-traffic controllers, other flight crews, maintenance teams, etc.). The lack of crew coordination can lead to confusion and poor decision making, eventually resulting in accidents. *Personal readiness* is the requirement for individuals to show up for work ready to perform at optimal levels. Personal-readiness failures occur when individuals fail to prepare physically or mentally for duty. These failures can lead to a pilot's rapid physical or mental fatigue during flight, with subsequent errors and accidents.

The third level of failure in the analysis system created by Shappell and Wiegmann, called *unsafe supervision*, comprises four categories: inadequate supervision, planned inappropriate operations, failure to correct a known problem, and supervisory violations.

Inadequate supervision is a failure of the supervisor to provide subordinates with the opportunity to succeed. The supervisor must provide guidance, training opportunities, leadership, motivation, and the proper role model(s). Sound professional guidance is an essential element of any successful organization. Many of the flight violations investigated have been traced to a lack of this element.

Planned inappropriate operations sometimes take place during tough flight scheduling periods. These operations can place pilots at unacceptable risk, jeopardize their rest periods, and ultimately adversely affect their performance. Improper flight crew pairing is another potentially dangerous issue—for example, when a young first officer is paired with a senior, dictatorial captain.

Failure to correct a known problem means that deficiencies among individuals, equipment, training, or other safety-related areas are known to the supervisor, who allows these problems to continue unabated. For instance, the supervisor's failure to correct a training schedule of a crew or the inappropriate behavior of a pilot creates an unsafe atmosphere and promotes the violation of rules. The failure to report known unsafe tendencies and initiate corrective action is another example of this unsafe supervision category.

Supervisory violations are rare events. They occur when supervisors willfully disregard existing rules and regulations. Such practices invariably set the stage for the tragic sequence of events that will follow.

Organizational influences are located in the fourth failure level of the HFACS. Fallible decisions of upper-level management directly affect supervisory practices, as well as the conditions and actions of flight crews. Potential failures sometimes may not be clearly visible. They are usually related to three managerial areas: resource management, organizational climate, and operational processes.

Resource management decisions in the air transportation industry normally are made to ensure flight safety and cost-effective operations. In times of prosperity, both objectives can be easily balanced and satisfied. In more difficult periods, safety and training are often the first to be cut. Excessive cost cutting could result in reduced funding for new equipment, purchasing of suboptimal equipment, poorly maintained workplaces, and the failure to correct known design flaws in existing equipment.

The result is a scenario involving unseasoned, underskilled pilots flying old and poorly maintained aircraft under the least desirable conditions and schedules.

Organizational climate can be viewed as the working atmosphere within the organization and consists of three basic elements: structure, policies, and culture. The *structure* defines the chain of command, delegation of authority, communication, and formal accountability for actions. *Policies* are official guidelines that direct management's decisions about such things as hiring and firing, promotion, retention, raises, sick leave, drug and alcohol abuse, overtime, accident investigations, and the use of safety equipment. *Culture* refers to the unofficial or unspoken rules, values, attitudes, beliefs, and customs of an organization.

Operational processes include operations, procedures, and oversight. This category refers to corporate decisions and rules that govern the everyday activities within an organization. It defines the establishment and use of standardized operating procedures and formal methods for maintaining checks and balances (oversight) between the workforce and management. The term *operations* defines operational tempo (pace), time-pressure actions, production quotas, incentives, measurement and appraisal, schedules, and deficient planning. *Procedures* consist of standards, objectives, documentation, and instructions. *Oversight* includes risk management and safety programs.

The HFACS method created by Shappell and Wiegmann allows for a profound flight accident analysis and bridges the gap between theory and practice in aviation safety securing. In addition, modern aviation practice indicates that flight crews must possess much more specific professional knowledge for safe operation of automated aircraft than for that of conventional aviation equipment. Systematic analysis of aviation incidents or accidents is impossible without adequate flight crew proficiency assessment. In this connection the Shappell-Wiegmann HFACS method must be supplemented by Kern's (1997) approach to professional flight crew formation. One core element of the Kern airmanship model is pilots' knowledge about themselves, their aircraft, their crewmates, their environment, and the degree of risk during the flying mission. This element can be added to failure level 2 (preconditions for unsafe acts) of the HFACS system and called "insufficient professional knowledge of operators."

Practical analysis of the accidents and incidents discussed in this book can begin with analysis of case 1, briefly described earlier in this chapter.

Automated airplane accident facts

To better prepare the reader for the accident analysis, it seems appropriate to look more closely at the flight history of the aircraft in question. It also seems useful at this point to briefly explain the nature of the automatic devices and functions mentioned in the accident description.

- The *go lever* is one of two identical small controls located on each thrust lever. This device allows the pilot on the ground before takeoff, as well as in flight, to immediately switch the aircraft automation into takeoff or go-around mode accompanied by a simultaneous corresponding increase of engine thrust and of the airplane wing angle of attack (which determines the degree of lift at a specific airspeed). This results in rapid aircraft climb.
- The *flight director* (FD) is one of the most important automated aircraft computer systems. It develops control signals for maintaining the flight

path parameters chosen by the pilot. FD control signals are provided for control bars of the primary flight display (PFD) and for the autopilot. In manual flight the pilot follows the deviations of the bars by manipulating the aircraft flight controls. In automatic flight the autopilot is controlled by the FD signals.

- The *alpha floor protection function* provides automatic increase of thrust to avoid aircraft stall (uncontrollable falling) in case the wing angle of attack increases to a critical value. When the alpha floor protection is activated, the symbol "THR-L" (thrust latch) is displayed on the flight mode annunciator (FMA), a special indicator located in the upper area of the A300-600 cockpit PFD.

The information presented below was retrieved from the aircraft digital flight data recorder (DFDR) and the cockpit voice recorder (CVR) and has been published on the Internet by the flight accident investigation agency (Sogame and Ladkin 1996).

11:14:05 Approximately at 1070 ft pressure altitude, the first officer inadvertently triggered a go lever. As a result, engine thrust increased; the aircraft developed a slight nose-up tendency and began to deviate above the instrument landing system (ILS) glide path. Speed also increased. In an attempt to recover the normal descent path, the first officer performed a nose-down operation by pushing the control wheel. However, the aircraft did not descend and, around 11:14:10, leveled off at approximately 1040 ft pressure altitude. The trim-of-horizontal-stabilizer (THS) position did not change from −5.3°.

11:14:09 The cockpit voice recorder (CVR) fixed an aural warning informing the crew about the automation landing capability change from landing mode to go-around mode.

11:14:10 The captain cautioned the first officer by saying, "You, you triggered the go lever," and the first officer acknowledged this, saying, "Yes, yes, yes, I touched it a little."

11:14:18 During level flight, both autopilots (APs) were engaged almost simultaneously. As the flight director (FD) was in go-around mode, the autopilots were also engaged in that mode.

11:14:20 Engaged autopilots began to move the THS from $-5.3°$ toward the nose-up direction, increasing the airplane nose-up tendency. The first officer tried to compensate this tendency by activating the pitch trim control switch on the control wheel. However, trimming of the THS using that switch is inhibited during engagement of the autopilot(s), so the first officer's actions had no effect.

11:14:23 The captain gave the first officer an order, saying "Push down, push it down, yeah." This is considered an instruction to push the control wheel down in order to correct the descent path, which had become too high.

11:14:26 The captain told the first officer, "You, that...disengage that throttle." This instruction seemed to indicate that the captain was telling the first officer to manually adjust the thrust by moving the throttle toward its idle position to correct the descent path, which had become too high.

11:14:30 Noticing that the flight mode annunciator (FMA) was displaying the go-around mode, the captain said to the first officer, "You, you are using the go-around mode," and then added, "It's OK, disengage again slowly, with your hand on..." There seems to be a possibility that in response to the captain's instruction, the first officer took some action to change from go-around

New Flight Operation Aspects Brought by Automation

mode to another pitch mode. That could have improved the situation, but no improvement occurred.

11:14:37 The THS moved automatically to $-12.30°$, strongly increasing the nose-up tendency.

11:14:39 The CVR recorded a sound that is assumed to indicate activation of the pitch trim control switch. As of **11:14:20** and **11:14:34**, however, this operation had no effect.

11:14:45 The captain again pointed out to the first officer: "It's now in go-around mode." The first officer answered, "Yes, sir." Although the first officer may have taken some action to switch from go-around mode to another mode, no mode change was actually made. At or around this point of time, the pitch angle and the angle of attack (AOA) increased and the speed decreased. To deal with this situation, the first officer slightly increased the engine thrust.

11:14:50 The sound of autopilot disengagement was recorded on the CVR.

11:14:51 The first officer said, "Sir, I still cannot push it down." With the aircraft pitch angle and AOA still increasing, and with speed decreasing, the crew continued its attempts to complete the approach. At approximately 570 ft pressure altitude, the thrust automatically increased suddenly, reaching its maximum at 11:15:03. This was caused by activation of the alpha floor protection function, which resulted when the AOA exceeded its threshold of 11.50° for the wing configuration of slats 30°/flaps 40° used for the approach.

11:14:58 The captain said, "I...that land mode?"

11:15:02 The first officer reported to the captain, "Sir, throttle latched again." Activation of the alpha floor function displayed a symbol of the thrust latch on the FMA. Owing to the thrust increase following activation of the alpha floor function at 11:14:57, the aircraft's speed and pitch angle increased; the aircraft stopped descending and began to climb.

11:15:03 The captain told the first officer that he would take over the controls. After doing so, the captain [now the pilot flying (PF)] pushed the control wheel to the forward limit, but the aircraft still continued to climb. Around this time the thrust levers were also temporarily retarded (held back), suggesting that the captain still intended to continue approach.

11:15:08 The captain said, "What's the matter with this?" The captain's response seemed to indicate that he was puzzled as to why the nose-up tendency was continuing, even though he had pushed the control wheel fully forward and decreased the engine thrust.

11:15:11 The captain again increased the thrust (which he had earlier reduced) while calling, "Go lever." At the same time, the CVR recorded the activation sound of the pitch trim control switch, and the digital flight data recorder (DFDR) recorded the movement of the THS in the nose-down direction. The captain said, "Damn it, how comes like this?" The captain's response appeared to express his bewilderment as to why the aircraft pitch angle was still increasing despite his actions to the contrary (pushing the control wheel fully forward and retarding the thrust levers). With the renewed increase in thrust, the aircraft began a steep climb with increasing pitch angle. Speed, which had earlier increased, began to decrease.

11:15:14 The first officer reported go-around maneuver to Nagoya Tower.

11:15:18 The sound indicating passage of the slats/flaps lever through the attached balk gates was recorded twice. According to normal go-around procedure, the slats/flaps lever should be moved from the 30°/40° position one step higher to 15°/20°. However, judging

New Flight Operation Aspects Brought by Automation 27

from the number of times the lever sound was recorded, it may have been moved beyond the 15°/20° position, perhaps to the even higher 15°/0° or 0/0 position. Later, at 11:15:27, a sound presumably indicating the lever's downward movement passing through the balk gate was recorded on the CVR. The DFDR also registered a record showing that the slats/flaps lever was set on the 15°/15° position at 11:15:28.

11:15:20 Both thrust levers were retarded almost simultaneously. At approximately 11:15:23, thrust lever 1 was retarded to the vicinity of its idle position and thrust lever 2 was retarded slightly. At approximately 11:15:27, both levers returned to almost full thrust positions.

11:15:21 The captain shouted, "Hey, if this goes on, it will stall!" Presumably this remark reflected the captain's shock either when he found that the aircraft was continuing to climb steeply with increasing pitch angle while reducing speed or when he noticed the position of the slats/flaps lever set by the first officer.

11:15:25 The stall warning began to sound because the AOA reached approximately 15° at 11:15:22, exceeding the threshold angle of 15° for the slats-extended configuration. It is surmised that around 11:15:25, the aircraft fell into a stall, yet continued to climb until reaching its highest point. Because of the thrust increase following activation of the alpha floor function, the aircraft's speed and pitch angle increased; the aircraft stopped descending and began to climb with an abnormally big angle of attack. That was determined to be the ultimate reason for the aircraft stall. The aircraft remained in a stall condition until impact.

11:15:26 The pitch angle of the aircraft reached the maximum angle of 52.56°(!). After reaching the highest point at approximately 1730 ft pressure altitude, with a pitch angle of 43.80°, the aircraft began to descend, while rolling and yawing greatly to the left and right. There are records showing that the crew took corrective actions by means of the ailerons and rudder during this period.

11:15:31 The thrust decreased temporarily. This was presumably caused by thrust surges that occurred in both engines.

11:15:34 From this point until just before the impact, the first officer shouted "Power" repeatedly. This was linked to his utterance of "Quick, push nose down" at 11:15:26 and is assumed to indicate a desire to increase thrust and thus recover lost speed.

11:15:35 The captain performed a nose-up operation using the control wheel. It is conjectured that before this moment the captain had been applying nose-down input to the elevator in order to decrease the pitch angle, but was now applying a nose-up input to the elevator to correct the decrease in pitch angle and the steep descent of the aircraft.

From the conditions in which the CVR and DFDR recordings ended, it was estimated that the aircraft crashed at approximately 11:15:45.

Accident analysis

Facts describing the accident make it possible to define activity deficiencies of each pilot individually as well as of both of them as a crew, and also in terms of improper management by their supervisors and by the airline's top leaders.

Unsafe acts: A crew's unsuccessful attempts to override the aircraft automation

To restore the manual pitch trim control of the aircraft, the crew did not switch from the go-around mode,

New Flight Operation Aspects Brought by Automation

activated by the first officer's inadvertent go-lever manipulation, to another pitch mode by pressing a button on the flight-control unit on the cockpit glareshield. The pilots might also have switched off both flight directors and autopilots in order to achieve fully manual control of the aircraft. Instead of those correct actions, both pilots tried to override the automation. These actions can be qualified as skill-based errors:

- To reduce a growing pitch-up tendency caused by automation switched into go-around mode, the first officer unsuccessfully tried to use the horizontal stabilizer trim manual control lever that could not be used in the go-around mode because of the airplane automation logics.

- The captain initially advised his first officer to push the control column, and then did that himself.

- Because of the crew decision error, it was too late to initiate a go-around procedure after the airplane became uncontrollable.

- In the last phase of the event, the crew again made a series of skill-based errors. Trying to save the flight and to find an exit from the unusual situation, the crew acted spontaneously, without any indication of understanding the situation.

Preconditions for unsafe acts: Lack of crew professional knowledge and resource management abilities
Neither the first officer nor the captain had sufficient knowledge to fully understand and to timely correct the abnormal situation caused by a single incorrect pilot finger movement. After the go lever was inadvertently activated, the crew's situational awareness decreased to an unsatisfactory level, because the crew did not know how to correct the plane's attitude and steer it in a safe

direction. Because of its lack of knowledge, the crew did not understand the new situation and could not effectively correct it. This precondition can be classified as insufficient professional operator knowledge.

The crew failed to use available resources to abort the emergency and prevent it from developing into a disaster. Apparently, both pilots commented only on what was actually occurring, without understanding why the aircraft was reacting in this way, and did not attempt even briefly to discuss how this difficult situation could be corrected. They also did not use the aircraft control system to complete the flight safely. By selecting another pitch mode, for example, they could have ensured a controlled climb or descent, as the flight-level change could have disarmed the go-around automation mode. However, considering the aircraft position relative to the runway, the only correct decision in that situation would have been to initiate a go-around procedure and then make another normal approach to land. The crew proved to be unable to discuss the situation and reach a decision together to properly use the aircraft equipment and follow the aerodrome navigation procedures. This precondition can be classified as substandard operator practice resulting from lack of crew professional knowledge and crew resource mismanagement.

Unsafe supervision: Insufficient crew training in aircraft automation operation and cockpit resource management

Both crew members were insufficiently trained in flight operations of the automated aircraft that they were flying. This is evident from the fact that neither the captain nor the first officer could fully understand the reason for the aircraft automation malfunction. Neither crew member had received adequate training in the use of all available resources in the cockpit to correct the first officer's skill-based error and to safely complete the flight. The flight crew supervisors directly responsible for both kinds of training had not provided the crew with the opportunity to succeed. A properly trained pilot does not perform inadvertent manipulation of the aircraft controls. A properly trained crew uses all available cockpit resources to overcome any abnormal situation. In that flight, neither of these conditions applied because of inadequate flight operation supervision.

Organizational influences: Company organizational climate

Airline organizational policy must contain areas of managerial activity devoted to reaching and maintaining high flight safety standards. A well-organized system of providing pilots with professional knowledge and skills needed for safe flight operations, combined with a reliable mechanism of crew proficiency assessment, is one of the most valuable tools in reaching an airline's economic and social goals. The case discussed above is connected directly with a low level of flight crew proficiency. The reason for this could lie in an airline organizational climate in which flight operations supervisors measured pilot proficiency merely on the basis of the pilot's total amount of flight time and previous transition training, and assigned commercial flight missions to an insufficiently trained crew, which, in the case cited above, resulted in an air disaster.

Disaster could have been prevented

There are at least four scenarios in which the incorrect action of the first officer could have been prevented from developing into a major accident.

32 Acquaintance with Aircraft Automation

1. The disaster could have been prevented by correct actions of the crew members as individual professionals.
   - The first officer could have prevented the accident if he had controlled his actions during the manual approach and had not touched the go lever, had been familiar with the aircraft automation features and the means of their operation indications and control, had understood what consequences would follow his inadvertent action, and had quickly assessed the situation and initiated a go-around maneuver.
   - The captain could have prevented the accident if he had known the aircraft automation features and the means of their operation indications and control, had been continuously aware of the flight situation and had actively controlled the flight by careful supervision, had not allowed the first officer to override the aircraft automation and had refrained from doing that himself, had accounted for the aircraft position and the automation indications after realizing the first officer's error, and had quickly commanded the first officer to make a go-around procedure or had initiated the maneuver himself.
2. The disaster could have been prevented by the flight crew acting in unison as a team of professionals to properly use all available resources to define the problem, discuss it, and make the correct decision to employ go-around mode.
3. The disaster could have been prevented if the flight division supervisors had provided the crew with an informative ground course and

sufficient flight and crew resource management training and had systematically tested the crew's proficiency.
4. The disaster could have been prevented if the airline's top management had established an organizational climate that provided pilots with the required level of professional training and had ensured that they were adequately trained and skilled before assigning them to commercial flight missions.

Pilot's Priority List: Automated Airplane Design and Operations Knowledge

Automation has brought significant changes to modern aviation. Long flight ranges, increased passenger and cargo capabilities, and previously unfathomable air navigation precision have dramatically expanded the commercial aviation industry. Not only for commercial aircraft and airline owners and business executives, but for millions of ordinary people as well, intercontinental flights have become routine events. Automated aircraft constitute the largest part of the leading world airline fleet. Pilots of automated airplanes are able to perform takeoffs and landings in very limited meteorological conditions.

Automation has changed the very essence of the pilot's profession. Although good airplane handling skills continue to be extremely important, stick-and-rudder proficiency is not sufficient for a contemporary pilot. The ability to make a highly computerized aircraft as understandable and obedient as the pilot's very first and still well-remembered airplane is another vital requirement for a safe, profitable, and pleasant flight today.

34 Acquaintance with Aircraft Automation

Knowledge is one of the most valuable qualities required in a modern aviator who operates or intends to operate automated aircraft in flight. Several areas of knowledge are vitally important for safe flying. Pilots must know themselves, their crewmates, the flight environment, and the flight mission. But to safely perform flight duties in an automated cockpit, each pilot must also have some basic knowledge about the aircraft being flown. The following chapters are devoted to attaining an important goal: expanding the reader's knowledge of automated airplanes. Automated aircraft design opens the door for a new list of pilot knowledge priorities.

2

Automated Aircraft Design

Automated aircraft, like all airplanes, consist of an airframe with conventionally located landing gears and flight-control surfaces (elevator, rudder, ailerons, flaps) used to control the aircraft flight path, and of a power plant creating the thrust needed for the flight and supplying the aircraft systems with the energy needed for their operation.

Automated Aircraft Controls

The elevator, ailerons, and rudder are the main devices that provide the flight path control of an airplane. In large conventional aircraft, cockpit flight controls (control columns, control wheels, and rudder pedals) are mechanically linked by cables, pulleys, and push rods to hydraulic actuators located in aircraft wing and tail areas. Those actuators cause movement of flight-control surfaces (elevator, ailerons, rudder). Conventional aircraft flight-control systems are equipped with elaborate devices of artificial feel to generate appropriate muscular feedback from flight-control surfaces to the pilot. Additional components providing trim compensation, mechanical stall warning devices, and devices for pushing the control columns to reduce the angle of attack are also included in conventional flight-control systems. To reduce takeoff and landing speeds, the aircraft are equipped with wing flaps and slats. Hydraulic or electric actuators usually cause movement of these aerodynamic surfaces.

The technological progress in aviation brought significant changes into aircraft flight-control systems.

Aircraft flight-control system development

The first automated aircraft were equipped with a system of flight-control powering similar to those used in conventional aircraft. Further improvement in aircraft economical efficiency and reliability required introduction of additional devices for increasing aircraft aerodynamic flexibility: a movable stabilizer, spoilers used for symmetrical or differential airlift reduction, and flaps with several aerodynamic slots that improved the airflow over the wing. However, realization of these goals was impeded by shortcomings inherent in most mechanical flight-control systems, such as a significant weight burden, the need for an efficient labor-consuming technical surveillance system, and a limited service life.

These shortcomings have been overcome by the introduction of a new electronic flight-control system that has, in principle, replaced heavy mechanical moving transmissions by lightweight electrical nonmoving circuits. In this system, termed *fly-by-wire flight control,* electric signals are sent through wires to set the hydraulic control actuators in motion. The signals are transmitted by switches operated by displacement of the pilot's cockpit control devices.

A typical flight-control system in a modern automated aircraft has two subsystems: primary and secondary flight controls. *Primary flight controls* include an elevator (often consisting of two symmetrical parts), ailerons, and a rudder. These flight-control surfaces are usually controlled by two sets of components that include control columns with control wheels or side-stick controllers, and rudder pedals. *Secondary flight controls* consist of a movable horizontal stabilizer, spoilers, slats, and flaps. Spoilers usually operate symmetrically as speedbrakes and differentially to assist ailerons for roll control.

To manually maintain the flight path parameters during flight evolutions, pilots of all types of aircraft must overcome the plane's tendency to undergo the short-term oscillations and long-term flight path deviations that occur as a result of the aerodynamic effects caused by the maneuvers mentioned above. For example, short-term bank oscillations normally occur in turns and are compensated for by opposite inputs of pilot flight controls. The tendency for long-term deviation during a turn must be compensated for by additional input from the pilot to trim the elevator by a special device that eliminates additional control surface loads.

The first automated aircraft control systems, although not as heavy as their predecessors, operated in the same way as conventional mechanical flight-control systems, with a degree of control force compensation. In those systems, all flight-control surface movements depended directly on increased input from the pilot.

In more advanced control systems the relationship between flight-control displacement and corresponding control surface movement was modulated for current speed, center-of-gravity position, and aircraft configuration. Increased pilot input was still required to compensate for short- and long-term flight path deviation tendencies. Thus, to maintain the flight path in a bank, the pilot had to produce bank command input; in a turn, to apply a back flight-control pressure; and in turbulent air, to assist the aircraft to maintain level flight. Flight-control systems of this type provided a good base for the design and production of two more sophisticated devices used in automated controls: dampers and automatic trims.

Aviation technology developments allowed for a significant increase in aircraft operational cruising speeds. This achievement was offset by a change in aircraft wing design that resulted in some reduction in roll stability. To compensate for this factor, more complex aircraft control systems have been designed. In these systems, which are still in widespread use in modern aviation, a short-term stabilization feedback provides quick calibrated movements of control surfaces. Special devices called *dampers* generate these movements. This type of compensation can be applied to all three aircraft axes (pitch, yaw, roll). Nevertheless, the most prevalent type of short-term compensation device is the yaw damper.

An additional long-term stabilization function called *automatic trim* is a feature of the most advanced fly-by-wire control systems. In this system, a computer device integrates pilot control inputs with feedback signals and provides continuous automatic trimming. This type of control system is used in commercial aircraft of the latest generation. There is only one instance in which the pilot trimming input is needed to maintain the flight path on these aircraft: change in airspeed. All other aircraft maneuvers, air turbulence reactions, aerodynamic effects, and configuration changes are compensated for automatically.

Basic principles of electronic flight control

Two kinds of electric signals are used in automatic control systems: analog and digital signals.

An *analog signal* is a continuous electrical copy of a mechanical movement; changes in this signal correlate with changes in movement of any mechanical control component. Analog signals are easy to develop from weak mechanical movements and following an amplifying process to turn them into more powerful movements. But electric signals of this class have one significant deficiency—they may be distorted during transmission and amplification. To eliminate this problem, analog signals are transformed into digital form.

Digital signals are more amenable to being copied, processed, amplified, and even saved. That's why high-quality electronic devices and all computers are digital. All automated aircraft use a special computer, called a *primary flight computer*, to process digital control signals along with other important information. The primary flight computer continuously receives information about actual flight conditions and factors in this information when processing the pilot's control signals. This computer function allows achievement of two important goals:

1. It facilitates the pilot's task of controlling flight in turbulent air or during changes in aircraft configuration. To compensate for perturbations caused by changes in airflow, the computer produces needed additional signals and mixes them into the flight-control signal.

2. It provides aircraft protection by preventing potentially dangerous flight situations such as stall or extremely high speed. As soon as a dangerous aerodynamic condition emerges, the computer produces an additional control input to compensate for its negative influence on the aircraft flight path.

The fly-by-wire control system operates as follows. A movement detector, called a *transducer*, detects the pilot's movement of a flight control in the cockpit and develops a corresponding electric analog signal. This analog signal is transmitted by electric wire to an electronic unit, which converts it into digital form, which is more appropriate for computer processing. Then the digital signal is fed into the primary flight computer. The computer processes the digital control signal together with other digital flight-relevant information signals and develops two new digital signals. One of these signals, after reconverting into analog form, is sent to the hydraulic actuator, which causes the control surface to move in the same way that the pilot moved the flight control in the cockpit. The other new digital signal is sent back to the cockpit flight control, providing the pilot with a muscular feedback identical to the feedback from a conventional flight control.

Automatic Flight-Control System

The aircraft automatic control system provides functions of full flight path parameter control. The most important components of this system are the primary flight computer (described above), the autopilot flight director system, and the autothrottle system.

Autopilot flight director system

As the term *autopilot flight director system* (AFDS) implies, this system provides two main functions:

1. Aircraft flight path automatic control
2. Flight path parameter calculation and indication

This system allows the flight crew to stabilize the aircraft around its center of gravity, to fly the aircraft on a required flight path, to acquire a new flight path, to perform an automatic landing, and to automatically select needed modes of the autothrottle system operation.

Automated Aircraft Design

Various automated aircraft have AFDS in their design and operations in a similar way, although specific devices and functions of this system may have different names. A typical AFDS comprises:

- Two or three identical flight-control computers (FCCs) that make all calculations needed for automatic flight path formation
- A unit that allows pilots to control whole-system operation: termed *flight-control unit* (FCU) for Airbus aircraft and *mode control panel* (MCP) for aircraft made by Boeing
- Hydraulic actuators that move aircraft pitch, roll, and yaw flight-control surfaces in accordance with the FCC commands
- Dynamometric rods and other devices used for specific functions of the system
- Flight mode annunciator (FMA) and flight director pitch and roll bar functions indication on the primary flight display (PFD), used by pilots to gauge automatic control system status and operation

Various separate controls and warning signals, which also belong to the AFDS system, are located in the cockpit. These include:

- The go levers [sometimes called *takeoff/go-around* (TO/GA) switches] used for immediate aircraft automation engagement in takeoff or go-around modes. These small levers are located on much larger engine thrust control levers. Some aircraft (e.g., Boeing 767) have two separate controls for takeoff (a button on the MCP) and for go-around (TO/GA switches on thrust levers) automation modes.
- The autopilot engagement switches or push-buttons on the glareshield.

- The autopilot disconnect pushbuttons on control wheels.
- The flight director (FD) and flight path vector (FPV) switches on the glareshield.
- The automatic landing status indication on the glareshield or on the pilots' instrument panels.

It must be noted that the AFDS has no separate autopilot or flight director block, box, or any similar devices. The autopilot functions as well as the flight director functions are performed by the whole set of the AFDS devices.

Autothrottle system

The *autothrottle system* is the second main part of the aircraft automatic control system. It is able to provide automatic control of engines' direct thrust during all phases of flight from initiating takeoff roll until the pilot deactivates it after landing. The main components, destination, and operation of the autothrottle system are as follows:

1. A computer (termed *thrust control computer* by Airbus and *thrust management computer* by Boeing) that performs all thrust calculations. The thrust computer receives and processes information from other computers engaged in aircraft control, as well as from the engines, radio altimeter, air data system, navigation system, corresponding switches, and pushbuttons. This computer then provides command signals to engine throttles via the autothrottle system actuator.

2. Autothrottle arming and engagement controls: arming lever and engagement pushbutton (Airbus) or one or two arming switches and autothrottle mode selection pushbuttons (Boeing). These controls allow pilots to prepare or activate the system for operation, or to switch off the whole autothrottle system or its left or right subsystem. The autothrottle mode selection

Automated Aircraft Design 45

pushbuttons allow manual autothrottle mode selection. Another way to select an autothrottle mode is to select it automatically via the flight management computer.

3. Go levers [or takeoff/go-around (TO/GA) switches] and autothrottle disconnect pushbuttons on both engine thrust control levers. Go levers are used to initiate takeoff thrust on some aircraft types and a go-around maneuver on all aircraft types. The autothrottle disconnect pushbuttons on engine throttle levers allow the pilot in certain flight situations to immediately disconnect the autothrottle (e.g., during a rejected takeoff).

4. A device used for preliminary setting of a needed thrust (a thrust rating panel or a thrust mode select panel on the cockpit instrument panel, or thrust setting pages of a computer display). This device allows the crew to select reference thrust modes for takeoff, go-around, climb, maximum continuous thrust, and cruise flight, as well as to select fixed and assumed temperature-derated reference thrust values.

5. A part (left side) of the flight mode annunciator (FMA) in the upper area of the primary flight display. This device indicates whether the autothrottle is engaged and what autothrottle mode is active.

6. Electromechanical components that include autothrottle system actuators, dynamometric rods, and throttle position detectors. They provide engine thrust control in accordance with the thrust computer commands.

The autothrottle system is normally operated in flight together with the autopilot flight director system (AFDS) while at least one autopilot is engaged. Before takeoff the autothrottle is activated and coupled to the AFDS by pressing a corresponding control (go lever or pushbutton). On earlier types of automated aircraft the autothrottle system can be operated in flight only when the aircraft is controlled by the autopilot. The latest aircraft types, such as Boeing 777, allow the autothrottle to operate in manually controlled flight without engaging the autopilot.

Flight envelope protection

The automatic flight-control system provides several protection functions that improve the safety of the aircraft by preventing the exceeding of its speed, angle of attack, and bank limits. The amount of automation authority and how it is activated in the flight envelope protection function vary with different types of automated aircraft.

Protection against extreme speed surges is one of the most common envelope protection functions. When the aircraft reaches a maximum speed, the automation issues commands to maintain that speed by reducing engine thrust (A310), or by producing a pitch-up force on the pilot's control column if the speed is increasing above the maximum limit (Boeing 777).

Automatic aircraft also have a protection function against speed reduction below a minimum value. One common feature that warns pilots of an approaching drop in speed or aircraft stall is a device called a "stick shaker." As soon as the speed reduces to the stall speed, this device produces control column vibrations to warn the crew about that dangerous condition.

In addition to the stick shaker, the automatic flight-control system of the latest aircraft types, such as Boeing 777, has a stall protection function that in certain conditions limits the airplane's ability to be trimmed in the nose-up direction. The pilot must apply a continuous aft column force to maintain the aircraft speed below a minimum maneuvering speed value. If the speed decreases to a near-stick-shaker activation value, the armed but—for some reason—disengaged autothrottle engages automatically and advances engine thrust control levers to maintain the minimum maneuvering speed.

A310 aircraft have a similar system called *alpha floor protection*. When an excessive angle-of-attack value is detected and the autothrottle system is armed, the thrust latch function is engaged. This action prevents engine thrust reduction and is provided to maintain a minimum aircraft speed higher than the stall speed.

There is another automatic flight safety protection function, directed toward recovery from windshear conditions encountered while in takeoff or go-around phases of flight. Engaging of takeoff or go-around mode provides the autopilot flight director system with windshear recovery guidance. During a manual flight, the flight director bars on the primary flight display show the pilot deviations of the flight controls needed to maintain a safe flight path. During automatic flight, the autopilot maintains a safe flight path. If the autothrottle is not armed, to maintain a required speed the pilot must manually advance the thrust levers to full thrust.

Boeing 777 aircraft have one more automatic flight protection function, called *roll envelope bank angle protection*. In cases of external environment disturbance, aircraft system failures, or inappropriate pilot actions, the system provides stabilization of roll control wheel input. When the airplane bank angle exceeds 35°, the control wheel force causes the airplane to roll back within 30° of bank.

Safe automatic flight-control conditions

The autopilot flight director and the autothrottle systems, if well understood and properly handled by flight crews and maintenance teams, can supply pilots with full authority to perform safe and efficient flights.

The working principles of all automatic flight-control and flight envelope protection functions and how these functions are applied must be clearly understood by pilots. This understanding is important to prevent very dangerous losses in situational awareness among flight crews and to prepare them for effective handling of automatic systems in any flight situation.

On the contrary, inadequate knowledge of automatic control systems and insufficient understanding of systems operations will inevitably lead to incorrect or inappropriate crew response during flight. Such operational errors of automated aircraft crew, independently or in combination with other organizational and managerial shortcomings, can significantly threaten flight safety and even lead to accidents.

Automated Aircraft Power Plant

To perform a flight, any airplane must have a source of thrust, needed for its initial acceleration during takeoff and for maintaining speed during flight. For normal operations, airplanes also require sources of electric energy and hydraulic and pneumatic pressure. All these forms of energy are provided by a power plant that is an important part of any airplane structure.

The power plant of automated aircraft consists of a propulsion system and an auxiliary power unit (APU). The power plant enables the aircraft to normally function both on the ground and in flight. The APU complements engines in supplying the aircraft with electricity and high-pressure air power. In normal operational conditions it is used on the ground and can be used either in flight or on the ground when an additional supply of energy is desirable.

Propulsion system

Aircraft engines make the propulsion system that creates thrust needed for airplane flight. Engines of modern aircraft also produce electric, hydraulic, and pneumatic power. The automated aircraft propulsion system normally consists of two, three, or four jet engines. Boeing 747-400 and Airbus 340 airplanes represent four-engine automated aircraft. MD-11 can be mentioned as a typical three-engine member of the automated aircraft family. These aircraft are acknowledged leaders in world aviation for their huge transportation capacity and long flight ranges. Automated aircraft types powered by two engines include A330, A320, A310, A300-600, Boeing 767, Boeing 757, and the last modifications of Boeing 737.

The Boeing 777 aircraft also has two engines and can be formally included in the third group of automated aircraft; however, in the author's opinion, this machine must be mentioned separately for its unique flight range combined with passenger and cargo capacities, and state-of-the-art design and equipment.

Modern long-range business airplanes are also competent members of the automated aircraft family because they are equipped with automatic flight-control devices and efficient engines similar to those mounted on large commercial aircraft.

Knowledge of the propulsion system and its correct operation by pilots is extremely important for safe flying of automated aircraft.

Automated aircraft engine basics

Engines of automated aircraft, which are expensive state-of-the-art machines, have low fuel and oil consumption, a high degree of reliability, long service life, and low noise levels. These properties are extremely important for airlines' economy, flight safety, and environment preservation. Competent operations of automated aircraft engines can be provided through pilots' understanding of engine design and operation.

A typical engine of automated aircraft is a dual-rotor axial flow turbofan of high compression and high bypass ratio. Its two rotors are mechanically independent. Rotor 1 consists of a fan, a low-pressure compressor section, and a low-pressure turbine section. A high-pressure compressor section and a high-pressure turbine section form rotor 2.

The fan creates the major part of the engine thrust; the rest of it is produced as a result of pushing back from the engine gaseous products of fuel burning. The thrust created by the fan can be reversed forward using a hydraulically or pneumatically activated thrust reverser.

Electric energy and hydraulic and pneumatic power are produced by special units mounted on the engine: electric generators, hydraulic pumps, and compressor air pressure valves.

Modern aviation engines are complex machines with carefully balanced parameters of operation at every phase of flight. For example, to start the engine, a special starter unit must initially accelerate rotor 2 to a definite speed of rotation (about 20 percent of its nominal speed). Then, at strictly defined moments, the fuel must be supplied and ignited. Not to exceed gas temperature limits that can destroy the engine turbine, the fuel flow must be slowly increased during further acceleration of the engine rotors. During engine start-up, pilots must carefully observe rotor rotation and oil pressure and gas temperature indication changes.

In processes of engine operation, when there is a wide-scale change in thrust caused by the pilot or the autothrottle, the engine parameters must be maintained within definite limits. Special engine-control electronic devices that provide optimum operation and engine structure protection during all phases of flight carry out most or all of those operations automatically. Only the most important actions, such as initiating engine start-up or shutdown, thrust reversal (engaging the thrust reverser levers), and thrust change in manual control, are performed by pilots.

Engine thrust controls
Manipulation of the forward thrust levers or the thrust reverser levers mounted on them provides engine thrust control. Forward thrust levers (called simply *thrust levers*) can be moved manually by pilots or automatically by the autothrottle. But thrust reverser levers can be moved only by the pilot when forward thrust levers are in idle position. This precaution is provided to avoid potentially catastrophic consequences of erroneous automatic thrust reverser activation in flight. Pilots are allowed to use the engine reverse thrust only during landing roll on the runway.

Engine parameter indication
Parameters that indicate (display) engine operation status are presented to pilots by a system of indicators located on the cockpit central instrument panel. The system can have a set of conventional instruments with pointers and round scales, or two electronic displays. Engine parameter indication on conventionally shaped instruments is used in some automated aircraft types (e.g., A300-600 and A310). In other automated aircraft, such as Boeing 767, the engine parameters are presented on cathode ray tube (CRT) displays. Display pictures of conventionally shaped round scales with pointers show the most critical parameters. Only important values of parameters are marked on scales, without any intermediate-value markings. Current parameter values are shown in digital windows beside the display images. Less important parameters are presented on displays in digital or graphic form. In the latest aircraft

models, such as Boeing 777, liquid crystal displays (LCD) are used instead of CRTs. Engine parameter indication on displays is part of a special engine indication–crew-alerting system. This system has two CRTs or LCDs.

The scope of engine parameters indicated by conventional instruments and by electronic displays is identical. A common principle of engine parameter indication is, the most important parameters are shown in the upper area of the instrument panel.

Conventional instruments are located in vertical rows, one row for each engine. Images of engine parameter indicators on displays are shown in a similar manner. The display area is divided into two segments, one of which is used for engine parameter indication. Engine parameters indicated on the central instrument panel by pointer indicators (from top to bottom) are as follows:

1. The *engine pressure ratio* (EPR) is a parameter that is proportional to the engine thrust. It shows the ratio of air pressures after and before the engine. An EPR reading multiplied by the frontal area of the engine gives its power value at the given moment. This parameter is controlled by the engine thrust lever position, which is manipulated by the pilot while in manual flight or by the autothrottle while in automatically controlled flight. Sometimes the EPR is not indicated, because it depends on the engine type used in a given aircraft. For example, in engines manufactured by Pratt and Whitney, this parameter is indicated, while in General Electric engines, the level of thrust is indicated by an N_1 reading.

2. The N_1 parameter indicates the rotation speed of the engine fan that creates the major part of engine thrust. Observation of this parameter is especially important for the flight crew during engine start-up or when flying in icing conditions. The fan rotation is

an important indicator of the whole-engine normal operation.

3. *Exhaust gas temperature* (EGT) is an extremely important parameter. The crew must carefully observe it during the whole flight. A crew error resulting in a failure to notice EGT growth above a critical limit during engine start-up can destroy the engine. Rapid EGT increase in flight can be a sign of engine surge or fire.

4. The N_2 parameter indicates rotation of engine rotor 2 with the air compressor and the gas turbine sections mounted together on the same shaft. This parameter is indicated as soon as engine start-up is initiated. Definite values of N_2 are used during engine start-up as triggers for other important actions and control checks, such as switching on the fuel supply by the pilot or automatic engine starter disconnect.

5. *Fuel flow* is the last engine parameter shown by a pointer indicator. This parameter is indicated during engine start-up after N_2 increase showing. Fuel flow is also an important engine operation parameter, allowing the pilot to assess normal engine operation according to the engine's fuel consumption and to predict fuel reserve at the end of flight.

Thrust reverser design and operation

The *engine thrust reverser* is an important device widely used on modern aircraft. It allows significant reduction of landing roll in normal conditions. It is also useful when the crew needs to reject takeoff because of an engine failure or other dangerous aircraft condition.

Engine thrust reversers can be used only on the ground when engine thrust levers are in idle position. To reverse the engines' thrust, the pilot must raise the thrust reverser levers mounted on the forward surface of the engine thrust-control levers. The airflow in each engine, accelerated by the engine's fan, is then directed forward by a hydraulically or pneumatically powered mechanism. Rotation of both engine rotors automatically increases to a value needed to produce a significant reverse thrust.

The effect of engine thrust reversal is most significant at high speeds immediately after aircraft touchdown. Manipulation of the engine thrust reverser lever at slow speeds can cause engine compressor blade damage and failure of other components, so this lever should be moved to the idle position after reduction of aircraft speed to a definite value. Normally, engine thrust reverse must not be used at speeds below 60 knots (nautical miles per hour).

Auxiliary power unit

The *auxiliary power unit* (APU) of an automated aircraft is a gas turbine engine driving an electric generator and an air compressor. It provides electric and pneumatic power for supplying aircraft systems and starting engines on the ground. During flight the APU can be used as an additional source of electrical and pneumatic power normally supplied by aircraft engines. All APU parameters during engine start-up and operation are controlled automatically. Only the most important APU functions are controlled manually.

Two APU controls are normally used by the crew: an APU start-up selector and an airbleed-control pushbutton. The selector is used for APU start-up and shutdown. The pushbutton is used to control the aircraft air supply from the APU. Electrical power supply from the APU is engaged automatically as soon as the unit start-up sequence is completed. The APU rotor speed and EGT are displayed in digital form on a central electronic instrument panel.

Aircraft Motion Control Errors

Modern commercial automated aircraft are heavy and equipped with powerful engines. Pilots must continuously control aircraft flight by monitoring operations of its control and propulsion systems, because these systems are sources of huge mechanical forces that cause all aircraft movement. Therefore, in case of failure of either the control or the propulsion system, even a small delay in crew response or corrective action can lead to a disaster.

Case 2: Lives lost on takeoff because of crew inattention

During a snowstorm on the morning of March 31, 1995, an A310 aircraft with a crew consisting of a captain, a first officer, and nine flight attendants, carrying 49 passengers on board, began a scheduled flight from Bucharest, Romania, to Brussels, Belgium.

The aircraft takeoff weight was well below the limit. Normally, when the aircraft weight is below maximum, takeoff thrust is reduced in such a way as to extend engine service life. Nevertheless, at this time the flight crew chose to apply maximum engine thrust for take off. This decision can be explained by considering the actual weather conditions. When the runway is covered with precipitation, friction between aircraft gear wheel tires and the runway surface is reduced, and pilots are advised to apply maximum engine thrust for takeoff. This precaution ensures that more runway length will be available for stopping the aircraft in case of aborted takeoff due to engine failure.

The first officer, serving as the pilot flying, performed the takeoff maneuver. He manually controlled the airplane with the autothrottle and flight director bar indicators engaged. Soon after takeoff, the aircraft was cleared by the air traffic control (ATC) tower to proceed in accordance with the flight plan to a waypoint called STJ. At 2000 ft of altitude, the first officer started a left turn directed to that waypoint.

At acceleration altitude the first officer, following indications of the flight director bars, reduced the airplane's pitch angle. The aircraft indicator speed increased to its flap retraction value, and the crew retracted the wing flaps. The aircraft was initially stabilized in a left turn with the following flight parameters: slats still extended in takeoff position of 15°, indicator speed 190 knots, and bank angle about 20°.

Usually, after an automated aircraft becomes accelerated to a required speed, the autothrottle reduces engine thrust from takeoff to the climb setting by moving the throttle levers back. Normally, during any change in automatic power plant thrust, all throttle levers move simultaneously and pass equal distances. But in this case the right throttle retracted only slightly. This moment can be considered as an emergency situation starting point.

The right throttle lever became stuck and could not be moved to the required new position. The right engine's thrust failed to reduce properly and remained above the climb rating for about 55 s in total. To compensate for an abundant thrust of the right engine and to prevent further increase in speed, the autothrottle system commanded the left engine to continuously reduce its thrust. After 42 s the left engine thrust was fully reduced to idle.

As a result of this thrust imbalance (i.e., right-left engine thrust asymmetry), the aircraft began to bank further to the left. In an attempt to counteract the banking tendency and to stabilize the bank angle near the required 20° value, the first officer made a right roll input to the control wheel. But the bank angle continued to increase.

When the bank angle increased to 45°, the first officer asked the captain to engage the autopilot. As the bank angle continued to increase, the pitch angle progressively decreased to 80° nose down. The left autopilot was engaged for only 1 s, and then it was disengaged as a result of a strong pitch force applied to the control column by the crew. Many large left and right control wheel inputs were also made by the crew, but they did not help regain control of the flight path. The aircraft crashed, killing all 49 passengers and 11 crew members.

Accident analysis

A mechanical failure was the starting point in the chain of events that led to the accident. The thrust-control malfunction was caused by excessive friction in the right throttle controls, which was attributed to unsatisfactory technical maintenance. The autothrottle actuator could not override this friction, and thus the right engine could not be automatically controlled.

A review of the aircraft maintenance logbook showed that similar malfunctions had occurred in this aircraft system before, but that no measures had been taken to improve maintenance to prevent the failure. After the accident, the digital flight data recorder (DFDR) information processing showed that abnormal operation of the same throttle lever had been registered in the previous flight from Abu Dhabi (in the United Arab Emirates) to Bucharest, but that the previous crew had not recorded this malfunction in the aircraft maintenance logbook.

High technology alone does not protect automated aircraft from failure caused by improper maintenance. Flight crew members usually notice such failure first. Unfortunately, in this case the crew proved to be unaware of the failure and lost the aircraft, together with its own and the passengers' lives.

Unsafe actions: Captain's routine violations and first officer's errors

Automated aircraft operations, with a crew consisting of only two pilots, require each pilot to perform a definite number of specific tasks. Only strict adherence to this rule can guarantee that all information available to the pilots concerning the flight will be clearly perceived and easily understood, and that all needed actions will be initiated and successfully completed in a timely manner.

Delayed response or inactivity, called *pilot incapacitation* (which can result from a heart attack or other suddenly emerging health problem), of either pilot may signal a potential disaster during flight. Prevention of pilot incapacitation is one of the most critical procedures for two-pilot crews, and pilots must be well trained in executing this procedure.

The flight operations manuals of all automated aircraft operated by two pilots specify that the primary task of the pilot not flying during takeoff phase is to monitor the engine parameters. Postflight analysis revealed that the captain did not follow this requirement. As the pilot not flying, he did not carefully observe the engine parameter indications, did not notice the abnormal autothrottle operation and thus did not immediately inform the first officer about this problem, and did not help him restore normal power plant operation by disconnecting the autothrottle and initiating manual control of engine thrust. These errors can be classified as *routine violations of flight operation procedures.*

The first officer, as the pilot flying, did not immediately inform the captain about difficulties in maintaining the flight path and did not ask him why the throttle was malfunctioning. Because of the captain's inattention or incapacitation, the first officer was the only person able to correct the situation. But the first officer did not try to find the reason for the throttle malfunction and correct this problem himself, without the captain's help. The first officer's actions can be classified as *skill-based errors*.

Another crew on a previous flight of the same airplane failed to record this throttle problem in the aircraft maintenance logbook. A special procedure requires flight crews to record in the aircraft maintenance logbook all failures that occur during the flight. The previous flight crew did not follow this requirement.

This unsafe action of the previous crew can be qualified as a *routine violation of aircraft operation requirements*.

Preconditions for unsafe actions: Crew resource mismanagement resulting in unsatisfactory coordination and communication

The crew resource management requires sufficient communication and coordination between crew members to ensure that they can function as a united team with open exchange of information and coordination of activity. Analysis of the case 2 flight revealed that the two pilots did not function as a team. They did not collaborate in detecting and understanding the problem and its cause or in developing a solution to this problem. The first communication between the pilots concerning adherence to the flight path was initiated only when the first officer told the captain that he was unable to reduce the bank angle and asked to engage the autopilot. Exchange of information between the two pilots was initiated only when the situation became catastrophic. Before this moment, only the first officer attempted to restore the normal aircraft control; however, his actions were insufficient. During most of the flight, the captain did not perform his duties adequately (i.e., acting as the pilot in command or as the pilot not flying). This failure became a precondition for unsafe actions that can be classified as *crew resource mismanagement*.

The crew members mentioned in case 2 were unaware of the autothrottle problem from the previous flight because the previous crew had not recorded it in the aircraft maintenance logbook. Such a notation might have helped the captain and the first officer to understand and cope with the problem in a timely manner.

Unsafe supervision: Insufficient crew training in aircraft automation operation and crew resource management

Prior to the flight (case 2), the captain and the first officer had not been trained or supervised sufficiently to enable them to perform their flight duties adequately, as they were unable to recognize the power plant thrust asymmetry, to solve the problem, and to maintain the aircraft flight path. Their crew resource management training was unsatisfactory because they were unable to act as a united team during the flight.

Organizational influences: Unsatisfactory organizational climate and company resource mismanagement

Organizational influences created conditions, namely, laxities in training requirements, that allowed the accident to occur.

An unsatisfactory organizational climate in the airline was manifested in tolerance to lax or low professional standards and poor discipline among maintenance personnel and flight crews. No actions had been taken to improve maintenance to correct similar throttle malfunctions on the same airplane following previous flights.

Thus, in case 2, the airline resources were not properly used to achieve the highest possible professional level of company flight crews. The unsatisfactory training or supervision of the flight crew members left them unprepared to recognize and handle the problem and to function as a team during the flight.

Accident could have been prevented
The accident might not have occurred if the captain, as the pilot not flying, had
- Continuously monitored engine parameters
- At the thrust reduction altitude, noticed a discrepancy between left and right engine thrust parameter indications
- Informed the first officer (the pilot flying) about the abnormality
- Helped the first officer restore the flight path and manually control the power plant

The accident could have been prevented by the first officer if he had taken the following steps immediately after noticing the problem:
- Fully moved the left throttle lever forward and made full control wheel and rudder right-side inputs to bring the airplane into level flight
- Informed the captain about the situation
- Disconnected the autothrottle by a pushbutton on any thrust lever
- Manually adjusted the thrust of both engines to maintain the flight path and used manual thrust control for the rest of the flight

The accident could have been prevented if the flight crew had acted as a well-coordinated team and had used the crew resource management principles for defining, discussing, and solving the problem encountered during takeoff.

The accident could have been prevented if the flight operations and aircraft maintenance supervisors had provided the flight crew with proper professional and crew resource management training and had required their subordinates to comply with aircraft maintenance and operational procedures.

The accident could have been prevented by the airline's top management if it had established and required from the airline employees a strict operational discipline. In that case the airplane would not have been cleared or released for flight with this defect, or the flight crew would have been prepared to handle the problem had it occurred again.

Pilot's Priority List: Remaining in Continuous Control of the Aircraft Flight

Description and analysis of the case 2 accident have emphasized the importance of flight crew preparedness, specifically the ability to always remain in control of the automated aircraft.

To do this, the crew would have to gather information about the aircraft, such as its mechanical condition and its flight path, and properly handle any emergencies during the entire flight mission.

A careful walk around the aircraft to study its external condition must precede every flight. The crew must then study the aircraft maintenance logbook from previous flights. The crew should also be briefed by maintenance personnel on all recent repairs made on the aircraft and on any future or planned repairs, deferred until later.

From the moment when the flight crew clears the aircraft for flight, following the captain's signature in the aircraft maintenance logbook, both pilots are responsible for timely and appropriate responses to all changes in aircraft status and flight path. During flight, any change in aircraft automation mode of operation or flight path parameters must be announced by one of the pilots and confirmed by the other pilot. The crew must be prepared to change an initially developed plan of actions in case of a failure. After the flight is completed, the crew must note any aircraft malfunctions or failures that occurred during the flight in the aircraft maintenance logbook.

During flight, the aircraft captain and the first officer must act as a well-coordinated team and exchange all information relevant to the aircraft's operation and the flight path parameters. They must be able to immediately detect any malfunction of or defect in any component of the aircraft, such as the engines, and to prevent further aggravation of the problem.

The pilot flying must immediately inform another pilot about any difficulties in controlling the airplane. The pilot not flying must carefully monitor the status of the engines and other systems, as well as flight path parameters, and immediately inform the pilot flying about any significant event concerning the flight status.

In addition, the captain of a given flight might be serving as either the pilot flying or the pilot not flying, but is always the captain and hence must continuously assess the flight situation, make proper decisions, and ensure that the aircraft crew remains in full control of the aircraft during all phases of the flight.

Part 2
Automated Aircraft Crew Working Environment

Part 2

3

Modern Flight Deck

Design of the flight crew cockpit, or the flight deck, has been one of the most important factors in flight safety improvement since the very first days of aviation. Every new step forward opened new possibilities and simultaneously brought new limitations.

The continuous technological progress, new procedures for solving flight operation problems, and flight crew working environment studies stimulated flight deck development. Flight safety, along with the cost-effectiveness ratio, has always been the prime factor in the process of the cockpit design optimization.

Automated Aircraft Cockpit Technology

The technology used in the aircraft cockpit developed as new technological achievements were implemented in aircraft. The technological progress has significantly changed aircraft systems: airframes, hydraulics, flight controls, engines, and flight decks. Over the years, the flight crew cockpit has received the most attention of all aircraft systems. The means of displaying flight information to the flight crew in the cockpit were also changed by the technological advances.

Flight deck development outline

The minor details of all modern aircraft flight decks are somewhat different, but there are two common features: face-forward orientation of crew seats and indication of flight-relevant information on cathode ray tubes (CRTs) or liquid crystal displays (LCDs). The latter is the main reason for calling these flight decks "glass cockpits." This term, albeit informal, is concise and clear and has been widely used in the aviation community.

The terms *flight deck* and *flight crew cockpit* are very close in meaning. The term *deck* usually means a raised separate room that has several crew seats, while the term *cockpit* generally means a smaller crew room for one or more people. In the context of this book, the difference between the two terms is not significant.

One of the first steps in modern aircraft flight crew cockpit design started from assessment of crew seat positioning within the flight deck. The face-forward crew seat design improved visibility outside the cockpit, along with working position comfort and ergonomics. It reduced the number of errors among the crew members because of their operational redundancy and ability to monitor each other's actions. These factors produced a great positive impact on safe flight operations and could easily justify the forward-facing cockpit layout. Another advantage of this design was the simplification of the human-machine dialog. Continuous and intensive research programs incorporating achievements of technological progress, as they became available for practical use, resulted in this new type of flight deck (Eduoard et al. 1981).

The increasing trend toward system digitization and the appearance of high-capacity microprocessors intensified the interest in the problems of improving flight deck layout. The introduction of computer technology into the flight deck became a reality. Numerous studies were made in mock-ups, investigating and evaluating flight-control and flight instrument layouts, systems panel ergonomics, and the various combinations of CRTs and conventional instruments that might be employed. These ergonomic and layout studies led to solution of the general flight deck architectural problem areas. However, the need also very rapidly became clear to improve system-operation methods and to reduce the number of instruments and indicators, leaving visible only those indications and readings needed when they were in use, obviously without any adverse effect on flight safety.

In the mid-1970s the world's leading aviation corporations began profound investigations into flight-control systems operation and monitoring, as well as basic studies on multifunction display technology. From the start, the studies were centered on a search for a modular and interactive display medium. This medium was found in the cathode ray tube, the only device at that time capable of satisfying all the environmental demands of aircraft operation. Parameters for computer application, the display logic, and the microprocessor programs were progressively defined as the result of multiple systems analyses.

To achieve the best possible results, the flight deck design has been the object of common work programs involving aircraft designers and flight crews from major aviation industry companies and airlines. The modern flight deck is the end result of a very long cooperative program between operators and constructors. There were two general goals in this cooperation: (1) The role of the flight crew had to be totally unchanged—it remained fully in command and was master of the decisionmaking process; and (2) whatever might be the evolution of operation and the constraints of routine, maneuver, and fuel economy in the future, the flight deck layout had to provide an improved means of achieving more efficient results together with a marked reduction in crew workload.

Electronic devices designed to increase flight crew awareness

Technology research programs were conducted in parallel with the systems operation studies to modify the aircraft controls and indicators. These programs were aimed at improving aircraft systems panels. As a result, it proved practical to integrate indicators in the control selectors, and also to integrate the selectors into the system synoptic diagrams. The human-machine interface was further improved by the adoption of the "lights out," or the dark cockpit, philosophy, under which no lights in the cockpit are visible as long as all is well. To simplify the evaluation of the level of an alarm or a warning, a color-coding system was also introduced.

Practical application of flight information processing and systems monitoring display demanded the creation of a cathode ray tube (CRT) suitable for functioning as an electronic indicator that could be used in airborne equipment. In early 1970s, a color CRT that satisfied those demands was created and improved to meet the flight deck high-altitude lighting environment conditions (Eduoard et al. 1981). A system comprising a symbol generator driving a CRT and a central alarm system was built for the U.S. National Aeronautics and Space Administration (NASA) in 1973. In the same year an electronic flight instrument panel, which contained six CRTs and was a forerunner of the present flight deck, was presented at the Le Bourget Exhibition in France.

A complete electronic flight-navigation system was constructed and tested from 1977 to 1979. The system consisted of the following equipment:

- Two electronic attitude director indicators
- Two electronic horizontal situation indicators
- Three data management systems and symbol generators to drive the four electronic indicators
- Two navigation system control and display units
- Two air data systems

To prove the electronic indication concepts and new instruments, the company Aérospatiale (in Toulouse, France) built a special flight simulator. The simulator's construction was based on the following criteria:

- A forward-facing crew seats layout
- Conventional flight controls providing the required visibility of the CRTs
- Electronic flight instruments mounted vertically, one above the other
- A central panel fitted with two CRT displays to provide warning information and systems and engine parameter monitoring, which later became identified as the electronic centralized aircraft monitor
- All primary information displayed on the CRTs, with a conventional instrumentation backup system

The simulator could also be used to test other new-technology equipment, such as illuminated pushbuttons, alphanumeric liquid crystal indicators, or digital frequency display systems.

Introduction of electronic indication systems into aircraft design and flight operations dramatically improved crews' abilities to be aware of any flight-relevant condition or event. More recently, more progressive liquid crystal displays (LCD) with better ergonomic and optical parameters have replaced CRT displays in automated aircraft cockpits. The flight deck of a modern aircraft has not been assembled simply from various pieces of hardware wrapped around a flight crew. It is the outcome of a multidisciplinary program applied to a vital aircraft organ to improve the safety and efficiency of its operation.

Glass cockpit concept

The expression "glass cockpit" means a cockpit where flight path parameters are presented on CRT displays. Although many airplanes have a radar indicator or other equipment with information displayed on a number of CRTs or LCDs, only those having the flight path information indicated on similar displays can be qualified as glass cockpit airplanes. The first attempt in aviation to display the flight path information on a CRT can be traced back to the late 1930s. The Sperry Company tested a CRT flight director, which could lead to a real breakthrough in instrument flying. However, there is no evidence of its commercial use (Underwood 1985).

The exact source of the term "glass cockpit" is unknown. It came into widespread use after the first commercial introduction of the CRT flight parameters indication concept in the Boeing 757 and Boeing 767 aircraft in the early 1980s. Both aircraft have identical cockpits, with an electronic flight instrument system (EFIS) that includes two electronic attitude director indicators (EADIs) and two electronic horizontal situation indicators (EHSIs) located in front of each pilot, and two engine indication and crew alerting system (EICAS) displays on the center instrument panel. Airbus models A300-600 and A310, built in Europe, can also be considered among the first glass cockpit aircraft. Their cockpit CRT displays have the same meanings, but with different names [e.g., a primary flight display (PFD) for attitude indication and, as mentioned above, the electronic centralized aircraft monitor (ECAM) for engine parameters and systems indication].

Aircraft building companies and aviation authorities were sometimes cautious in introducing new technology into the flight crew cockpit. For example, although Boeing wanted to be innovative, it didn't want any crew transition problems. Pilot acceptance of new indication devices was a critical factor. For this reason, the first-generation EFIS displays looked like electronically shaped conventional instruments. The U.S. Federal Aviation Administration (FAA) was also cautious because it had to certify new airplanes.

One of the truly innovative achievements in the use of EFIS was initially proposed as only a secondary CRT display mode. This was the alternative horizontal situation indicator (HSI) presentation where the traditional 360° compass rose would be switched to a look-ahead ±35° variable-scale sector map showing required track, waypoints, navigation aids, airports, and the like with a weather radar data overlay. Pilots quickly adapted to the sector HSI display. These days it is difficult to see a pilot continuously flying with the HSI display in the 360° compass rose mode.

There are two more CRTs in the Boeing 757/767, A310, and other aircraft cockpits. They are installed for the dual control of the flight management system (FMS). Devices with these displays are called *control display units* (CDUs). In the flight decks of the latest aircraft, such as Boeing 777, three CDU displays are used to control, along with FMS, other information and communication aircraft systems.

Although there have been a number of changes since the late 1970s in the ways in which the information is presented to aircraft crews, the main features of the electronic flight deck are much the same as they were in the early times of the glass cockpit development.

Aircraft System Status Electronic Indication

The *aircraft system status electronic indication* is an important device that allows the flight crew to obtain information about the operation of other airplane systems. This device is especially helpful because no crew member is responsible for this kind of aircraft operation. It provides pilots with status information and enhanced ability to control the airplane engines, fuel, and flight-control, hydraulic, electrical, pneumatic, and other systems. The first good examples of such systems were Boeing 757/767 EICAS and Airbus 310 ECAM systems. These two systems can be considered predecessors of other similar systems widely used in all automated aircraft.

Engine indication and crew alerting system (EICAS)

The Boeing 757/767 engine indication and crew alerting system (EICAS) was designed to display engine parameters in normal operations and to alert flight crews to abnormal situations. The EICAS also can show information about the airplane systems status that is used by the crew to determine the airplane's ability to perform a flight. On the ground the EICAS supplies maintenance personnel with a variety of airplane systems data.

The core elements of the EICAS are two computers that receive and process inputs from engine and airplane system sensors. These computers operate independently and continually monitor each other. The information from the computers is displayed on two CRTs, located one above another in the cockpit on the central instrument panel. The space on both displays is vertically divided in two areas. One side area shows images similar to those created by conventional pointer instruments that represent engine parameters. Another area is used for text messages informing pilots about aircraft system operational conditions.

Upper EICAS display

The left side of the upper EICAS display is used for abnormal-situation alerts. Crew alert messages are shown in this area to indicate, in abbreviated text form, information about abnormal operation of airplane engines and other systems. The crew-alerting EICAS segment continually monitors all aircraft systems. A crew-alerting message is displayed on the upper CRT whenever a fault occurs in any system. In addition to the upper CRT messages, some crew alerts are also indicated by aural signals and warning or caution lights.

Abnormal-situation messages

Three categories of alert messages alert the flight crew to abnormal situations: warnings, cautions, and advisories.

Warnings are the most important category of crew alert messages. They indicate urgent operational or system conditions that require immediate corrective actions from the crew. Engine or APU (auxiliary power unit) fire and rapid airplane decompression at a high altitude are examples of warnings.

Cautions are less urgent, but are also important time-critical alerts. These messages require timely corrective actions. Low oil level and engine overheat can be mentioned as examples of cautions.

Advisories are the least urgent alert messages. They indicate operational or system conditions that can be corrected by the crew at an appropriate time. Excessive compartment temperature or "door lock open" indications are examples of advisory messages.

EICAS messages that are not considered crew alerts, but identify systems faults affecting the aircraft departure, are called *status messages*.

Before departure, the flight crew must use two documents in case any EICAS alert or status message is displayed. These documents are called the *dispatch deviation guide* and the *minimum equipment list*. They establish requirements for the equipment needed to fly the airplane.

Lower EICAS display

Additional engine parameters are shown in the left area of the lower EICAS display. This information includes oil pressure, oil temperature, oil quantity, and engine rotor vibrations.

The lower EICAS display is also used for status message indication. The flight crew can turn it into status page display. In this case the following airplane systems parameters are shown in the upper left corner of the left display:

- Hydraulic liquid quantity for all three hydraulic systems
- Hydraulic pressure in hydraulic systems
- Auxiliary power unit (APU) parameters: exhaust gas temperature, rotor rounds per minute, and oil quantity
- Oxygen pressure in the crew oxygen system
- Temperature in the bulk cargo compartment (this compartment can be used for animal transportation)

The upper and medium right side of the lower display can show various status messages describing airplane system conditions that the crew must know before and during flight. If the displayed status conditions are within established limits, these messages are not necessarily identified by either alert messages or discrete lights on system panels.

A diagram of the airplane flight-control positions appears in the left lower corner of the status display. It shows the rudder, ailerons, and elevator positions and is used by the flight crew before departure to check the flight control's normal operation.

In the latest EICAS modifications, an indicator of main gear wheel brake temperatures is shown in the lower right corner of the left display. The temperature of every wheel brake is represented by a numbered figure (1, 2, 3, etc.); the higher the temperature, the larger the corresponding figure. This indication is very helpful for timely initiation of a brake cooling procedure after landing in hot weather, when short flights are performed (e.g., airfield flight training), or in case of a rejected takeoff, as well as in other cases where wheel brakes are intensively used.

During flight, some indications shown in the lower display, such as rotor 2 rotation speed and fuel flow, are not continually used by the crew and can be periodically switched off, as they are relatively insignificant. In this case the lower EICAS display becomes blank, and additional information about engine operation is moved to the upper display to be shown in abbreviated form.

Electronic centralized aircraft monitor (ECAM)

The Airbus family of automated aircraft, such as Airbus 310 and 300-600, have an electronic centralized aircraft monitor (ECAM) that has the same destination as the EICAS, but is designed in a different way. The ECAM system assists the flight crew by providing information on normal and abnormal flight operations. The system includes

- Two flight warning computers
- System data analog/digital converter
- Two symbol generator units
- Two CRT displays
- ECAM control panel
- Master warning and master caution lights
- Audio interface
- Maintenance control panel

The *flight warning computers* (FWCs) acquire and process operational data from aircraft systems and develop warning, caution, and status functions. The most critical failure signals are supplied to both FWCs. Other signals are divided between the computers, with the possibility for data exchange through a crosstalk bus. Each FWC has the capacity to acquire and process all critical and non-critical failure signals. Flight warning computers generate various warning and control signals supplied to the automatic flight-control system, stick shaker, master warning lights, and other aircraft systems.

The *system data analog/digital converter* (SDAC) acquires various data and converts them into a digital format necessary for further data processing.

Symbol generator units (SGUs) process outputs from FWC, SDAC, and other aircraft systems and develop visual information and messages presented on displays. Each SGU is capable of driving both ECAM displays.

The ECAM control panel allows the crew to select:

- A system synoptic display showing the chosen system diagram and its main units
- The PLAYBACK (recall) function of a failure sequence
- The STATUS function showing actual systems condition
- Removal of a warning message presentation from the ECAM display

Master warning (red) and master caution (amber) lights provide visual warning information to alert the crew without identification of the failed system. There is one red warning light and one amber warning light for each pilot on the cockpit glareshield. Flashing of red warning lights requires an immediate crew action. Red or amber warning light activation is accompanied by a corresponding aural warning.

The AUDIO interface includes two loudspeakers, a warning cancellation pushbutton, and a cancellation switch.

The MAINTENANCE control panel is used on the ground for aircraft maintenance and troubleshooting.

The ECAM system information is presented on two CRT displays. The left ECAM display is the message display and operates in three modes: RELATED TO FAILURE mode, STATUS mode, and MEMO mode. This display presents warning messages and status and memo information. The right display is the system display. It presents aircraft system synoptics.

Left ECAM display

In normal conditions, when both CRTs are in operation, the left display presents one of four pages. RELATED TO FAILURE modes are presented on pages 1 and 2. Page 3 presents the STATUS mode, and the MEMO mode is presented on page 4.

Page 1 presents independent or primary failures. This presentation requires immediate application of a corresponding procedure. As soon as the FWC detects a warning, the left CRT automatically shows page 1 with a warning message written in a box under its system title. Crew actions needed to isolate the failure are written above the warning message. The crew can manually display page 2 while page 1 is displayed at any given moment by pressing the CLEAR pushbutton on the ECAM control panel.

Page 2 presents a secondary failure, if any, caused by a primary failure. Secondary failures that were caused by a primary one are selected on page 2 on the same CRT after all actions required for the primary failure isolation have been completed. An asterisk precedes the name of a system that is affected by a secondary failure. After secondary failures have been isolated, page 3 is displayed on the left CRT by pressing the CLEAR pushbutton. This page can also be displayed at any time by pressing the STATUS pushbutton on the ECAM control panel. The STATUS mode presented on page 3 allows the crew to know the operational status of the airplane.

Page 3 presents the aircraft status information, including the following:

- Operational summary of the airplane's condition
- Explanation of possible AUTOLAND capability downgrading
- Indications of the airplane's status following failures that affect the flight
- General contents and organization of flight-relevant data, such as emergency procedures, landing capability, flight path limitations of speed and altitude, deferred procedures, lost systems or functions, and other important information

Page 4 is the MEMO page. It is the basic mode page of the left ECAM CRT. The MEMO mode is selected automatically when no other mode is engaged. It can be canceled manually by pressing the STATUS or RECALL pushbutton. This mode is canceled automatically as soon as another mode is engaged. The MEMO page presents

- A reminder of systems or functions that are temporarily used under normal operations
- The total amount of fuel on board
- The total air temperature
- The fuel tank temperature
- The airplane's center-of-gravity value
- The airplane's gross weight
- STATUS indication, if there is any status information
- Indication of takeoff or landing warning message inhibition

As soon as the crew members acknowledge a warning message, they must perform all actions written on pages 1 and 2, then read and consider the information shown on page 3. The last page to be displayed on the left ECAM CRT after failure isolation is page 4.

Right ECAM display

The right display is the system display. It presents aircraft system synoptics. In normal display conditions, when both CRTs are in operation, the right display presents 1 of 14 airplane system schematics:

1. Engine starting
2. Engine operation
3. Hydraulic systems
4. Electric alternate current system
5. Electric direct current system

6. Airbleed system
7. Air condition system
8. Cabin pressure system
9. Fuel system
10. Auxiliary power unit
11. Flight-control system
12. Doors
13. Wheels
14. Cruise page showing engine parameters and airplane cabin environment conditions

The system of presentation symbols and color coding is widely used on the right ECAM display. Every class of system device is depicted by special symbols. For example, pumps are represented by rectangles, circles are used to show valves, and so on. If a system is in normal operation, it is shown in green color. The schematic of systems unusable because of failure is amber, while systems that are not in use are not shown at all.

Numerical values or analog indications are green when in the correct range. When a warning threshold of a parameter is reached, its numerical value and the corresponding pointer initially become amber, and as long as the abnormality aggravates to a more dangerous condition, they turn red.

ECAM operation modes

The ECAM system operates in four modes. Three modes are automatic and one mode is manual.

The *normal mode* is automatically engaged only when no other mode is engaged. When in normal mode, the left CRT reminds the crew by showing the MEMO page, which in systems used temporarily, such as engine anti-ice or the SEAT BELTS ON sign, operates at a given moment.

The right CRT shows a system page that is most appropriate to the current flight phase, in accordance with the ECAM inner logic. The master warning and master caution lights are off, and all pushbutton switches on the ECAM control panel are also extinguished.

The *manual mode* is operative when no failure is detected. The crew can engage the manual mode at any time. This mode is disengaged as soon as the advisory mode or the failure mode is automatically engaged. When in manual mode, a system page selected by the crew, except the CRUISE page, is shown on the right ECAM CRT.

The *advisory mode* engages when a parameter drifts out of its normal range. To attract flight crew attention, the right CRT shows the relevant system page well before reaching the warning level. The word ADV (in white) flashes in the upper area of the display beside the system title. The parameter symbol that triggered the advisory mode is pulsing smoothly. All local warning lights on airplane system control panels are off. This mode is inhibited on the ground and in flight until the slats retract.

The *failure mode* engages as soon as a warning is detected by the FWC. A message associated with the failure is shown on the left ECAM CRT. The right CRT shows the page associated with the failure (if any) to allow the crew to crosscheck the affected part of the system and to follow results of corrective actions. This mode has priority over all others' modes. Nevertheless, the flight crew can call any other system page using the ECAM control panel.

Isolation of Aircraft System Failures

The EICAS and ECAM text messages that inform the flight crew about airplane system failures are identical to messages shown in the *quick reference handbook* (QRH), a book with flaps that lists the most probable scenarios for airplane system failure and recommendations for crew actions needed to reduce the consequences of failure and to successfully complete the flight.

As soon as an EICAS message is identified, the Boeing 757/767 airplane crew must take the QRH, open it on a page with the title similar to that of the EICAS message, and start the failure isolating procedure by reading QRH lines from top to bottom and performing the recommended corrective actions step by step.

The A310 ECAM system allows the crew to isolate failures by reading the needed information from the left ECAM display and performing corrective actions without using the QRH. The ECAM text is identical to that of the QRH, and in most cases there is no need to use the QRH for failure isolation. The A310 QRH is used only for airplane performance calculations in various operational conditions and as backup for the ECAM system in case of its failure.

Multifunctional Display

The Boeing 757/767 EICAS and Airbus 310 ECAM are powerful tools for the flight crew to monitor and correct all important airplane systems operation conditions in a timely manner. Boeing 777 aircraft are also equipped with an EICAS similar to that used in the 767 aircraft model. The 777 EICAS indicates engine and airplane system parameters and provides a display of crew-alerting messages, system status, system synoptics, and maintenance information on two LCDs located one above another on the central instrument panel.

A new function in multifunctional display (MFD) format is also performed by the left inner, right inner, and central lower Boeing 777 EICAS displays. The display

Modern Flight Deck 87

select panel (DSP) on the right side of the cockpit glareshield controls the MFD format. In case of a display failure, the display system automatically reconfigures or can be manually reconfigured to compensate for the fault.

The MFD function allows selection of any of the three displays for presentation of various information, such as

- Navigation display
- Status page
- Secondary engine parameters EICAS display
- Synoptics of electrical, hydraulic, fuel, air, door, gear, and flight-control systems
- Communication pages
- Electronic checklist

Most of these displayed information classes were described earlier in this book. But two new items have introduced the latest achievements in the flight crew cockpit design and technology: communication pages and the electronic checklist.

Communication pages

Communication pages on the MFD are used to control data-link features that allow the crew to request and receive, through available satellite and conventional radio communication channels, various data concerning the flight, such as flight routing, weather, operational advisories, and commercial information. The data-link messages can be printed on the flight deck printer. Incoming messages are annunciated on the upper EICAS display.

Electronic checklist

The Boeing 777 *electronic checklist* is another new function of aircraft automation that makes possible a paperless operation of this airplane.

A list of mandatory operations needed at definite stages of normal flight preparation and flight performance, called a *checklist*, is an important part of every airplane flight operations manual. To facilitate the use of the normal operations checklist, it is often printed on a separate piece of paper, cardboard plate, or plastic frame with small windows for every operation, and is located in an easily reachable place of the cockpit. Every time the crew starts a new flight, the normal checklist is used to assure procedure-completed performance from pre–engine start-up cockpit preparation through airplane secure procedure after flight.

In the event of an airplane equipment failure, another reference document, the *abnormal checklist,* is used. This checklist allows the crew to correct the failure by strict step-by-step adherence to an abnormal procedure shown in the checklist. Abnormal checklists are contained in the airplane QRH or in another similar document.

Boeing 777 aircraft designers allowed the flight crew members to check normal flight operations without having to find and follow entries in printed (hardcopy) checklists using their fingers. For each phase of flight, the crew easily calls a section of the airplane normal checklist on a multifunctional display (usually on the central lower MFD screen) by pressing the corresponding button on the display select panel (DSP) once. Flight phase checklist section sequencing is automatic. All needed operations are shown on the MFD screen in white lettering. As soon as an operation has been completed, its lettering becomes green. After the whole checklist section is complete, information to that effect is shown in the display lower area. If the CHECKLIST button on the DSP is pressed again, the checklist section display is removed.

In case of a failure, annunciated in text form on the upper EICAS display, an abnormal checklist associated with the failure is automatically displayed on the MFD after a crew member presses the same CHECKLIST button. The crew must isolate the failure in the usual read-and-do manner. The system automatically changes the colors of closed-loop items as soon as they are completed. Pilots must use a cursor to mark completion of several non-closed-loop items, much as would be done using a conventional computer.

If a failure or a situation, like a window crack or a required overweight landing, cannot be annunciated on the EICAS, the crew calls an abnormal checklist menu and selects the needed section with the cursor.

If an electronic checklist failure occurs, the crew can use printed checklists contained in the airplane's QRH, which contain all information needed for normal and abnormal operations.

Aircraft-Status Monitoring Errors

The status of some aircraft systems may change during flight because of a failure. If the flight crew members do not factor in the new aircraft status, flight conditions can become difficult for them, especially in the presence of other negative factors. In such a situation, the probability of crew error increases dramatically, especially when the crew members have inadequate professional training.

The case described below did not lead to an accident. At the very last moment the captain managed to restore full control of the flight. But this case shows that even a minor automation failure can seriously aggravate a difficult flight situation if the crew is not fully qualified.

Case 3: Lost airspeed on approach to land

On May 22, 1997, an A310 aircraft crew performed a scheduled flight from Jakarta (in Indonesia) to Dhaka (in Bangladesh). The captain was pilot flying (PF) and the first officer was pilot not flying (PNF). The autopilot was engaged after takeoff. There were no abnormalities after departure from Jakarta until the level flight. Some time after reaching flight level 360, the airplane autothrottle system failed. For the rest of the flight, the crew manually operated the engine thrust.

It was a dark night when the crew started to descend for approach to land. Thunderstorms were observed in the vicinity of Dhaka airport. The crew commenced an instrument VOR (very-high-frequency omnidirectional radio range) approach to runway 14 with the autopilot engaged and the failed autothrust system. The airplane's gross weight was 118 metric tons, with a required airspeed of 144 knots for flap 20 configuration used for initial approach. During the initial approach, while descending to 1500 ft, and shortly after reaching this altitude, the captain controlled the airplane, avoiding cumulonimbus clouds by activating the autopilot in the HEADING SELECT mode with manual engine thrust control.

During the final portion of descent, the engine thrust levers were in idle position. Soon after reaching 1500 ft altitude, the captain started to perform a turn in a level flight to the final approach course. Distracted by the weather radar observation and trying to control the aircraft to catch the final course, the captain failed to notice the sharp reduction in airspeed and to control it by manipulating the engine thrust levers accordingly. The first officer remained inactive during the approach, did not monitor the flight parameters, and did not inform the captain about the speed loss. The speed reduced to 122 knots, which was well below the lowest selectable speed value for that given aircraft's weight and configuration. The autopilot continued to maintain the altitude by increasing the aircraft's pitch.

As soon as the airplane's pitch increased to approximately 20°, the stick shaker automatically activated, showing approaching stall flight conditions. At that moment the captain noticed the airspeed loss and aggressively increased engines thrust to 95 percent of N_1. The airspeed increased to a normal value, and the airplane restored its normal flight. Further approach and landing were completed without any deviations.

Additional information:

The crew members performed their first flight to the Dhaka airport.

The policy of that particular airline required all captain pilots to have flown as independent observers at least once to any airport that they had never serviced (i.e., at which they had not landed their aircraft in the past). In violation of this requirement, neither the captain nor the first officer had ever flown to the Dhaka aerodrome before the incident.

The crew members' flight experience at the moment of the incident was as follows:

- The captain was a 41-year-old male. He had a total flight time of 8911 h; total flight time on A310 aircraft, 1617 h; and flight time as a captain of A310 aircraft, 695 h.

- The first officer was a 47-year-old male. He had a total flight time of 12,719 h, but his A310 aircraft first-officer flight time record was only 339 h. Before the transition training for A310 aircraft, he was a captain of a conventional aircraft operated by a crew with a navigator and a flight engineer.

The A310 autothrottle failure is clearly indicated on both the pilot primary flight displays and on the ECAM. It can be removed neither manually nor automatically at any time during the flight.

There were no navigation data for the Dhaka airport in the airplane flight management computer (FMC) database, and the crew constructed the approach maneuver in the FMC by manually entering the required data into it through the control display unit (CDU) from the available flight documentation.

The incident was discovered in the process of recorded flight data analysis, which was a part of the airline's flight safety control policy.

Weather conditions at the Dhaka airport at the time of the incident were as follows: wind 090° at 15 knots, visibility 6 km, overcast cumulonimbus clouds with cloud base 1500 ft, temperature 32°C, aerodrome pressure QNH 1004 hectopascals (hPa).

Incident analysis

As soon as a pilot has recorded the first several hundreds of flight hours on automated aircraft, flying this machine (the A310) may seem to be an easy task. Alas, as in any machine, any system in an airplane can fail. Sometimes this happens even with automated aircraft systems. If pilots do not account for the actual aircraft status, the failure can suddenly change a smooth, easily performed flight into a serious challenge for the crew. The time deficit emerges, and it inevitably leads to crew errors. But the flight crew must have professional reserves to overcome the difficulties imposed by the failure and to successfully complete the flight.

This case illustrates a similar situation. The crew members at this time were lucky to overcome the problem, which was to a great extent a result of their own and other people's unsafe actions.

Preconditions for unsafe actions: Substandard practice of operations, including crew resource mismanagement and personal readiness

The crew members were making an instrument VOR approach at night, to an unfamiliar airport, with the failed aircraft autothrottle system.

Preparation for the approach was complicated by the absence of information on airport conditions in the aircraft database. During the approach, thunderstorm weather conditions in the vicinity of the airport required additional crew attention to avoid cumulonimbus clouds while following the approach route and maintaining a safe altitude.

The VOR approach performed by the crew is a relatively complex maneuver in comparison with the *instrument landing system* (ILS) approach. The VOR approach is also known as an *instrument nonprecision approach*. This approach requires from the crew careful preparation for the approach and good crew coordination.

When performing the approach, neither the captain nor the first officer demonstrated the requisite crew resource management ability. The captain tried to perform all operations alone. He did not urge the first officer to actively perform the pilot-not-flying duties, which included attention to the autothrottle system failure. The first officer was not ready to properly perform his flight duties in this flight. He did not have the ability to take part in the crew teamwork and was complacent in taking a passive observer role.

In this flight, an example of substandard practice of operations resulting from inadequate crew resource mismanagement was expressed by both crew members, especially by the first officer's lack of personal readiness to perform the flight.

Unsafe actions: Perceptual and skill-based crew errors and the first officer's routine violation

The captain's perceptual error was the immediate cause of the incident. During the approach, the captain (PF) was overloaded with flight-control tasks. In nighttime flight conditions, making an approach to an unfamiliar airport, the captain simultaneously had to follow the approach route, avoid cumulonimbus clouds, catch and maintain the final approach course, and manually maintain aircraft airspeed. The amount of information that needed to be processed proved to be above the captain's perceptual abilities, and he lost control of the airspeed.

The first officer's (PNF) skill-based error was the prime factor among those that created unsatisfactory working conditions for the captain. He did not take account of the autothrottle failure and did not help the captain by carefully monitoring the flight parameters, nor did he notice the airspeed reduction and thus did not alert the captain to this dangerous situation.

The first officer's routine violation was his main error. Operation procedures and the very design of all automated aircraft controlled by two-pilot crews strictly require that the pilot not flying continually monitor all flight parameters, inform the pilot flying about any deviation, and be ready to help the pilot flying to control the airplane. But the first officer, who was a former captain of a conventional aircraft, routinely violated the requirement and chose the role of a passive observer in that flight. This dangerous attitude of the first officer can perhaps be explained by overconfidence; for many years all members of his former crews had worked under him, and he had not worked under the supervision of others on any flight.

The errors of both pilots, aggravated by the first officer's violation, were the immediate causes of potential disaster, which could have severely threatened flight safety.

Unsafe supervision: Supervisory violation and inappropriate operation planning

Unsafe flight operations supervision resulted in assignment of insufficiently trained crew to perform the required flight mission tasks; the tasks were beyond the professional abilities of this crew.

The flight operations managers violated company policy by assigning the crew to a flight mission to an unfamiliar airport (Dhaka) without checking as to whether the captain had ever visited that airport. The crew members performed their first flight to the Dhaka airport, which they had never seen before, even as observers.

The flight operations managers had planned an inappropriate operation by assigning the flight mission to a crew that did not have satisfactory training in crew resource management. The crew pairing could also have had a negative effect on flight safety. The first officer, who was 6 years older than the captain, had much less A310 aircraft flight experience and was not a good choice to accompany the captain under those specific conditions.

Organizational influences: Unsatisfactory organizational process, resource management, and organizational climate

Unsatisfactory organization of flight operations by the airline was the reason for the inappropriate mission assignment to an unqualified flight crew.

Poor company resource management was the reason that the flight crew lacked sufficient crew resource management (CRM) training and that needed navigation data were absent from the airplane FMC database.

A low level of discipline in the company was confirmed by unsatisfactory duty performance by managers who did not provide the crew with the possibility to succeed as well as by a crew that did not report the incident as soon as possible.

Incident could have been prevented
Neither the captain nor the first officer performed their professional flight duties properly, nor did their supervisors.

The incident could be prevented if the captain had done the following:

- Before departure, carefully assessed his crew's readiness to safely complete the flight and insisted that another crew be assigned to the flight, or that an instructor pilot be included in the crew
- Organized crew activity during the approach in such a way as to secure the flight path and maintain adequate flight parameters
- Urged the first officer to properly perform pilot-not-flying duties

The incident could be prevented if the first officer had

- Not violated the requirement that the pilot not flying perform a head-down flight while carefully monitoring all flight-relevant indications
- Informed, in a timely manner, the captain about the reduction in airspeed and helped the captain to maintain the flight path parameters

Automated aircraft require that their crews act as a well-organized team. The incident could have been prevented if the crew members had carefully prepared their aircraft and themselves for the approach, had shown adequate crew resource management ability, and had demonstrated good crew coordination during the approach. In this case the first officer would have immediately noticed all aircraft flight parameter deviations and quickly informed the captain about them. The captain would have had more time to assess the flight situation and to make correct decisions.

The incident could have been prevented if the flight operations managers had not assigned the flight mission to a crew that was insufficiently trained in crew resource management; whose captain was totally unfamiliar with the airport, never having visited it even as an observer; and whose first officer had only modest experience in flying this type of airplane.

The airline's top executives could have prevented the incident and significantly reduced the flight safety risk if they had managed the company resources in a way that secured sufficient flight crew CRM training. An acceptable level of supervisory discipline established by the top management would have prevented the violation of company rules, specifically that an unprepared crew had been assigned to the flight. Supplying the airplane database with the needed navigation information could also have improved the flight crew's preparation for the landing approach and reduced the risk of crew error.

Pilot's Priority List: Aircraft System Status Considerations

Monitoring the actual aircraft status, especially during important phases of flight, can help prevent pilot error in unexpected or difficult flight conditions.

Many aviation disasters could have been prevented if airline executives and other personnel had acted in a professional and disciplined manner. Regretfully, this ideal condition is not always observed. Nevertheless, in a rare but still possible chain of unwanted events, errors, and violations, added to aircraft system failures, there is one final factor that can prevent a disaster. It is called *crew proficiency*.

98 Automated Aircraft Crew Working Environment

In every flight, and especially in flight on automated aircraft, the captain, together with the first officer, must organize and perform proficient monitoring and assessment of the aircraft system's operational condition. This action can help pilots cope with difficult flight situations and thus safely complete the flight.

4

Flight Path Parameter Electronic Indications

4

To control the airplane, pilots use information models of the flight created in their own brains. They know what the airplane flight path parameters should be to achieve goals set in the flight plan and maintain those parameters using the aircraft controls in the cockpit. Pilots' actions are directed to minimizing differences between ideal mental flight models and images of real flights created in their brains according to their perceptions of the flight parameters. The major part of information perceived by pilots in actual flight environments is visual and does not require any special devices to enhance perception by the pilot: the space and objects outside the airplane, the horizon, and the position of airplane controls in the cockpit. Other types of flight-relevant information, such as airplane systems operation data or flight path parameters to be perceived by the pilot, require visual interpretation assisted by special instrumentation or indicators. These devices were introduced in the aviation industry to better indicate aircraft system conditions or flight path parameters, and research and development (R&D) to improve their design has continued for some time.

Electronic Flight Instrument System

Automatic aircraft are equipped with an *electronic flight instrument system* (EFIS), which is designed to display the flight path parameters during the entire flight. Although the EFIS equipment used on various types and models of automated aircraft may vary slightly, the basic design, structure, and functions of all EFIS systems are similar. The world commercial aviation flight safety data record for the 1990s proved that pilot knowledge of the operational principles of flight automation and correct interpretation of EFIS indications are usually the most important factors in eliminating flight crew errors and preventing modern aircraft malfunctions and flight accidents.

Flight path parameters represented on the EFIS are divided into two groups: actual and selected. *Actual parameters* describe the flight path condition at a given point in time. *Selected parameters* are desired flight path values chosen by the pilot or automatically.

Color coding is widely used in modern flight decks to enhance the efficiency of pilot perception of flight-relevant information. For example, six colors are used in the Boeing 777 aircraft cockpit to indicate flight-relevant information and conditions:

White	Displays present status and range scales
Green	Shows dynamic conditions
Magenta or pink	Shows command information, pointers, symbols, fly-to conditions

Blue or cyan	Indicates inactive or background information
Amber or yellow	Corresponds to cautions, faults, or flags indication
Red	Used for warnings

The typical automated aircraft EFIS includes two primary flight displays (PFDs), two navigation displays (NDs), and two EFIS control panels. One set of the three items is mounted in the aircraft cockpit for each pilot, together with a number of supplementary electronic equipment and switches.

Primary Flight Display

The most critical flight path parameter indication in an automated aircraft cockpit is provided on a specific display called the *primary flight display* (PFD). Although the picture layout and number of indicated parameters of the PFDs of various automated aircraft may differ, the Boeing 777 PFD has practically all features that can be found in other airplanes.

Two identical PFDs are mounted in the cockpit, on both left and right instrument panels. Each PFD presents all parameters needed for flight path control. The parameters indicated are airspeed, vertical speed, barometric altitude, radio altitude, attitude, steering indications, flight path position indications, time-critical warnings, ILS station identification, heading and track indications, approach minimums, and marker beacon indication.

Airspeed indications

Speed is the aircraft life-critical flight path parameter. At any given time a speed vector of the aircraft longitudinal movement relative to the ambient-air mass can be represented by its two components: airspeed and vertical speed. Because these two components are so important, their values are displayed on separate gauges to pilots in all airplanes. In the glass cockpit the actual airspeed is displayed on the left side of the PFD in a tape form and in a digital window. The tape moves along the left side of the PFD in accordance with the speed changes. The actual speed values are shown on the tape scale against a pointer formed by the speed window and in digital form within the window. The selected speed can be displayed in two ways: in digital form above the speed tape picture or by a "bug" shaped as a wide horizontal arrow on the tape scale.

The pilot's ability to predict the future state of the airplane is very important for flight safety. This task is facilitated by a feature called the *airspeed vector*, which is built into the electronic speed indication. The airspeed vector is presented as a vertical arrow that shows, along the cutting edge of the tape, the predicted airspeed value after 10 s. If the arrow is directed up, the pilot knows that the speed is increasing, and vice versa. This indication is especially valuable during flight in turbulent air, on final approach to land, and in windshear conditions. This enhanced ability to predict airspeed changes enables the crew to make timely thrust and pitch attitude corrections and thus maintain a safe flight.

Several important airspeed values are displayed along the right edge of the speed tape.

Takeoff speeds (velocities) V_1 (decision speed), V_R (rotation speed), and V_2 (safe takeoff speed) are displayed during takeoff. The V_1 speed helps pilots decide whether to reject takeoff or to continue the flight in case of a major failure during the takeoff roll. The V_R value tells pilots when to rotate the airplane by the elevator to increase its wing angle of attack, which leads to the airplane's liftoff from the runway. The V_2 speed provides the airplane with sufficient controllability during the initial climb after liftoff. This speed must be maintained until an acceleration altitude is reached, where the airplane airspeed is increased, allowing the wing flaps and slats to retract. These takeoff speed values depend on the airplane takeoff weight, the runway length and elevation, ambient-air temperature, and other factors. Pilots calculate them before each takeoff using special tables or the aircraft flight management computer (FMC).

Another group of airspeed values indicated on the PFD are the flap maneuvering speeds for the corresponding flap positions. These values are the minimal airspeeds allowed to maintain definite wing flap settings. They are indicated by figures corresponding to flap settings in degrees located on the right edge of the PFD speed tape.

Before each approach to land, the crew calculates a *landing reference airspeed*; this is a speed value that strictly corresponds to an intended landing configuration of the airplane and must be maintained during final approach before landing. The abbreviation REF beside the speed tape marks this speed.

Edges of color bands along the right side of the speed tape indicate maximum and minimum speeds allowed to maintain flight. The *minimum maneuvering speed*, represented by a yellow band along the right edge of the speed tape, indicates the maneuver speed margin to the stick shaker activation, which warns pilots of approaching stall flight conditions. The *maximum maneuvering speed* is represented on the PFD in the same manner; it indicates a buffet maneuver margin and may be displayed during operation at high altitude and relatively high gross weights. The *minimum speed* indicates the airspeed at which the stick shaker activates. The *maximum speed* is the maximum speed allowed for a given airplane configuration. A black-and-red band along the speed tape shows both of these dangerous speed zones.

In addition to the usual airspeed indication in knots (nautical miles per an hour) or kilometers per an hour, all modern aircraft have one more way of indicating speed. During flight at high altitudes, a *Mach number* is used for airspeed indication, showing the proportion between actual airplane airspeed and the sound speed in the ambient air at a given altitude. When approaching the speed of sound, the airplane wing lift and drag values change significantly. The Mach number shows the pilot an actual aerodynamic condition of the airplane wing operation. When the current Mach number is greater than 0.40, it is displayed below the speed tape in digital form.

The vertical speed is not displayed on all glass cockpit PFDs. Often this parameter is indicated by a conventionally shaped pointer instrument located beside the PFD, as on the A310 aircraft. But in the Boeing 777 cockpit the vertical speed is displayed by an image of a pointer indicator located on the right side of the PFD. The pointer shows climb or descent as on a conventional indicator. In addition to a vertical speed value indicated by the pointer, the climb/descent value is digitally displayed when it is greater than 400 ft/min (feet per minute). It is displayed above the indicator with positive vertical speed or below the indicator with negative vertical speed. If the vertical speed (V/S) pitch mode is selected in the mode-control panel (MCP) of the airplane autopilot flight director system (AFDS), a bug on the indicator scale shows the selected vertical speed value.

Barometric and radio altitude indications

The aircraft altitude is measured in one of two ways: above one of the assumed theoretical barometric levels and above a ground surface flown over at a given moment of time. The aircraft barometric altitude is displayed on the altitude tape along the right side of the PFD in a way similar to the speed display. Digital altitude values are always displayed in feet and, if selected on the EFIS (electronic flight instrument system) control panel, are also displayed in meters. Selected altitude is displayed above the altitude tape in digital form and as a bug on the tape. The digital selected altitude value is shown in a box when the aircraft approaches the altitude.

Barometric altitude values can be displayed on the PFD in one of three ways relative to one of three barometric references chosen by the crew:

1. A standard atmospheric pressure altitude is an altitude above the isobaric surface of 1013.2 hPa (hectopascals). This altitude reference is marked on the PFD as STD.
2. An altitude above a hypothetic sea level at a given point on the Earth's surface. The assumed reference air pressure surface is marked as QNH.
3. An altitude above a given point on the Earth's surface calculated by a ground atmospheric pressure, marked on the PFD as QFE.

Pilots select values of QNH and QFE through the EFIS control panel. These values can be preselected in flight while STD altitude indications are displayed, and activated when needed, for example, at transition flight level during descent before the approach to land.

Radio altitude is indicated on the PFD when its value is below 2500 ft above ground level. The radio altitude is displayed in digital form in the bottom center of the attitude indication area. The indication color becomes amber when the aircraft descends below radio altitude minima.

Attitude indications

Directions of airplane movements in flight and often the airplane flight stability depend on the attitude that can be expressed through the airplane bank and pitch values; hence the attitude indication is very important for correct airplane control in flight and for the very safety of the flight. When discussing an airplane movement within a short period of time (minutes or even seconds), the Earth's spherical form can be neglected. In this case the horizon can be represented as a visible or imagined line on an endless plane. This assumption allows calculation of the airplane bank angle as an angle between the airplane's lateral axis and the horizon. Similarly, the airplane *pitch angle* can be defined as an angle between the airplane's longitudinal axis and the horizon.

The airplane attitude referenced to the horizon is displayed on the PFD through airplane pitch and bank indications. The principle used for these indications is similar to that used in conventional attitude indicators. A line dividing light and dark parts of the indicator's background imitates the horizon. The horizon line moves in accordance with airplane pitch and bank changes. A symbol of an airplane is displayed in the central area of the indicator; this symbol does not move. The airplane attitude in terms of pitch and bank is represented by the horizon line position located on the PFD relative to the airplane symbol.

Bank angle is represented by the attitude of the airplane symbol against the horizon line and pitch scale. A pointer in the upper area of the indicator indicates bank angle values. When the bank angle increases to a significant value that can negatively affect the flight conditions, the bank indication changes color to attract the pilot's attention.

Pitch attitude is indicated by pitch values shown on a movable vertical pitch scale between the airplane symbol and the horizon imitation line. A pitch limit indicator, shown in the form of two symmetrical horizontal brushes, is displayed at low speeds when the flaps are up and at all times when the flaps are down.

Slip and skid conditions are also indicated on the PFD by a small rectangle under the bank angle pointer.

Steering indications

Steering indicators show the desired flight-control inputs needed for flight path changes that provide adherence to the intended flight path or prevent collision with other aircraft. Steering is indicated in one of two ways, according to different sources of information: (1) the airplane flight-control commands calculated by the autopilot flight director system (AFDS), based on comparison of the airplane's desired and actual flight paths, and (2) automatically received and processed data on the positions of other airplanes.

The desired flight path parameters are entered into the AFDS either automatically from the flight management computer (FMC) or manually by the pilot through the mode-control panel (MCP).

Two flight director bars located in the central PFD area display the flight director steering indications. The vertical bar displays lateral steering indications, while the horizontal bar displays the pitch steering indications. The indications are active on the PFD when the associated flight director switch on the MCP is in the ON position.

To provide safe separations between the airplane and other aircraft, the traffic alert and collision avoidance system (TCAS) calculates so-called traffic advisories and resolution advisories. The TCAS shows other aircraft located at safe distances as white symbols on the PFD. When the distance to another aircraft reduces to an established value, this is indicated by a yellow traffic target symbol. As soon as the distance reduces to a smaller value, the color of the target mark turns red. And finally, when the distance becomes dangerous, a resolution advisory is displayed in the PFD attitude area in the form of trapeze-shaped lines. To avoid a collision, the pilot must change the airplane attitude in such a way that the center of the airplane symbol is moved outside the trapeze. A red band emerges simultaneously along the vertical speed scale. The aircraft vertical speed needed for a safe separation is provided when the vertical speed pointer is positioned outside the band advisory.

Flight path position indications

The PFD can indicate locations of actual and desired flight path positions relative to the airplane by means of flight path vector and flight path angle functions. The *flight path vector* (FPV) indicates the actual airplane flight path angle and drift angle. The FPV symbol is represented by a circle with two lateral short lines (wings) and one vertical short line (tail). The symbol shows the angle value between the flight path and the horizon in the vertical plane. The drift angle is represented by the perpendicular distance from the centerline of the pitch scale to the symbol in the lateral plane. Another symbol, represented by two symmetrical pairs of short horizontal lines, indicates a selected value of the vertical *flight path angle* (FPA). FPV and FPA indications are very useful during visual approach to land or when no other flight path guidance is available. These functions are activated on each PFD by a FPV switch on the corresponding EFIS control panel and by selection of FPA on the MCP.

During an instrumental landing system (ILS) approach to land, the desired flight path location is shown on the PFD by the ILS indication. This indication shows airplane deviations from the ILS glideslope and localizer beams radiated by corresponding ground radio facilities. In the beginning of the ILS approach the glideslope pointer and its scale appear on the right side of the PFD, and the localizer pointer and the localizer scale appear on the bottom of the PFD.

Time-critical warnings

Time-critical warnings appear on the PFD if the flight crew must immediately act to prevent an emergency situation from developing into a disaster. These warnings appear in a lower area of the PFD between the attitude display and the heading/track compass rose. They are shown in large red capital letters.

- The WINDSHEAR warning appears when a windshear condition is detected.

- The PULL UP warning appears when the ground proximity warning system (GPWS) has determined that the barometric descent rate or the degree of radio altitude decrease is excessive, or that the look-ahead terrain warning is active.

- The ENG FAIL warning appears when an engine has failed while the speed is between 65 knots and slightly less than V_1 (the decision speed).

Other PFD indications

Other PFD indications appear in different zones of the PFD.

1. The ILS station identification or frequency, course, and other parameters associated with the ILS distance measuring equipment (DME) reading appear in the upper left corner of the attitude display area.

2. Heading and track indications are presented in a section of the compass rose shown in the bottom area of the PFD. A pointer at the top of the compass rose displays a current heading. A bug representing the MCP selected heading is displayed on the outside of the compass rose. Another bug representing the MCP selected track is displayed on the inside of the rose. The MCP reference switch (HDG/TRK) determines whether the heading or track is displayed. An annotation (also termed *annunciation*) on the left side of the compass rose shows whether a heading or a track is selected. The current magnetic or true reference of heading or track selection is displayed in the right side of the compass rose. The line drawn from the invisible center perpendicular to the rose edge shows the current airplane track.

3. Approach minima are defined by radio altitude or barometric altitude. For each PFD, minimum values are selected by knobs on the corresponding EFIS control panel. The minima are shown at the bottom left side of the altitude scale.

4. The marker beacon indication appears in the PFD upper right corner.

5. The aircraft automation operation modes are indicated in the upper area of the PFD by a special device: the flight mode annunciator (FMA).

While controlling the aircraft flight, the pilot needs two classes of flight-relevant information: data describing the aircraft flight parameters at a given moment, and data allowing the pilot to assess the aircraft position in airspace relative to the route waypoints, airports, navigation facilities, and other airplanes. The information presented on the PFD reflects mainly parameters of the airplane flight at a given moment. Another electronic device, the navigation display, is used to ensure a continuous flight path indication and observe the environment around the airplane.

Navigation Display

The *navigation display* (ND) is a device that provides the flight crew with the aircraft progress and horizontal situation indication. There are two identical navigation displays in the cockpit, one ND for each pilot. The ND has its specific ways, or modes, of operation to expand its ability to present information.

Modes of ND operation

There are four modes of ND operation: approach, VOR [very-high-frequency omnidirectional radio (range)], plan, and map. The *approach* and *VOR* modes display track, heading, wind speed, and direction with VOR navigation or ILS approach information. These two modes are used during definite, relatively short periods of flight. The *plan* mode is used to view the active route in a step-by-step manner. The *map* mode is used during most phases of flight.

The *approach* mode displays on the ND localizer and glideslope information that can be useful during an ILS approach with or without the flight director indication. In this mode the ILS data, course, and DME distance are presented.

The *VOR* mode is used when the crew needs to intercept and follow a definite VOR radial: a bearing, calculated as an angle between magnetic north meridian of the VOR station location and a given direction from the VOR station. This mode can be used while on route or during a so-called nonprecision instrument approach to land. In the VOR mode the VOR tuning information and the course, DME distance, and to/from indication are displayed.

The *plan* mode is used for the route entry control during preflight preparation or when the route is changed in flight. For example, after receiving an ATC (air traffic control) clearance to fly directly to a waypoint, the crew must check the new routing before approval of the change execution in the flight management system (FMS). If the distance to a new waypoint is beyond the ND map range, the only way to perform the check is to use the plan mode, which allows successive manifestation of all waypoints along the route.

The *map* mode shows the airplane position relative to the flight route, which is displayed within the ND. An isosceles triangle top in the lower middle zone of the ND shows the airplane's current position. During flight the route image shown on the display moves from top to bottom. Information displayed on the ND in the map mode includes selected and current track, selected and current heading, position trend vector, range to selected altitude, map range scale, ground speed, true airspeed, wind direction and speed, next-waypoint distance, waypoint estimated time of arrival, and selected navigation data points.

The map mode display expanded abilities

During flight the ND in the map mode can supply the crew with a wide scope of information needed for safe and efficient flight navigation. Coupled with other aircraft equipment, this mode allows display of three safety-critical data groups:

1. Information about dangerous weather phenomena such as cumulus clouds, thunderstorms, and even turbulent-air zones
2. Pictures of the Earth surface flown over, with information about terrain heights
3. Symbols representing other aircraft in the area, with information on their relative movement

To help pilots in routine short-term planning of the flight, two features are introduced in the map mode: a position trend vector and range indication to a selected altitude. The *position trend vector* shows a calculated position of the airplane in the nearest future moments. The end of a dashed line starting from the aircraft's current position represents the calculated future position of the aircraft. The line consists of three segments, showing predicted aircraft positions after 30-, 60-, and 90-s intervals. When the airplane is in turn, the vector assumes an arc configuration with a radius defined by the airplane's bank angle and ground speed. A selected range of the ND indication determines the number of visible trend segments. For the range of 10 nautical miles (nmi), only one segment is displayed; for the 20-nmi range, two segments; and for a range greater than 20 nmi, all three segments. The *range to a selected altitude* is displayed while the aircraft is in either climb or descent mode. A wide green arc crossing the selected track line represents the range. The range value depends on the airplane's ground and vertical speeds, wing configuration, and landing gear position.

The ND can be operated in expanded or center map submodes. The expanded submode shows a front sector 80° wide, while the center submode provides the crew with navigation information at equal distances in all directions concentric to the aircraft symbol located in the

ND. Displays of additional navigation facilities, waypoints, airports, route progress, and position data are available in both expanded and center submodes.

Avoiding navigation display errors

The picture shown on the ND is very informative. Nevertheless, pilots must understand that the displayed airplane route and position data are not an immediate reflection of actual facts, but merely computer-calculated estimates. In most cases these estimates are continually corrected by ground radio navigation facilities such as VOR and DME, or by the global positioning system (GPS). But sometimes these corrections are not reliable or are even impossible, and the ND map picture does not reflect the actual situation. To avoid undesirable errors in air navigation, the flight crew must always compare the map display data with so-called "raw data" received directly from available navigation facilities and displayed in the cockpit by corresponding indicators showing actual VOR radials, DME distances, and nondirectional (radio) beacon (NDB) bearings.

The information indicated on the ND in a form of simple lines and text would be difficult to distinguish and perceive. To make the navigation information presentation more convenient and reliable, numerous symbols are used. These symbols indicate on the ND heading, track, wind direction and speed, radio navigation aids, and other objects and conditions. The TCAS (traffic alert and collision avoidance system) data as well as the radar and look-ahead information, indicated on the ND, are also represented in symbolic form. To perceive and correctly understand the ND information, pilots must know the meanings of those symbols.

The mandatory amount of information presented to pilots on the PFD and the ND is defined automatically in accordance with the stage of flight. To achieve better situational awareness and to make correct decisions, pilots can change, within definite limits, the information presented or add more flight-relevant information to the displays. This can be done through an electronic flight instrument system control panel.

Electronic Flight Instrument System Control Panel

The *electronic flight instrument system* (EFIS) control panel controls the amount and types of information presented on the PFD and the ND. EFIS information is supplied to PFDs and NDs through two EFIS control panels, one for each pilot. The panels are located in front of pilot seats in the cockpit on left and right sides of the glareshield. Pilots using their respective EFIS control panels can independently control the corresponding display options, modes, and range for the left and right PFD and ND.

Parameter indication controlled on PFD

Each pilot can control, through the EFIS panel, various flight parameter indication settings on the PFD:

- Landing minima are set by the MINS (minimums) reference selector, which has outer, middle, and inner knobs. The outer rotating knob selects RADIO or BARO (barometric) altitude as the PFD minimums reference. The inner rotating knob allows the pilot to adjust the PFD radio or barometric minimums altitude. The inner push knob resets the PFD minimums alert display and blanks the minimums display.
- The flight path vector (FPV) switch, when pushed, displays the flight path vector.

- Altitude indication in meters is provided by pushing the MTRS (meters) switch.

- Standard barometric setting (29.92 in Hg or 1013 hPa) for the PFD altitude scale reference is displayed by pushing the inner STD knob of the BARO switch. If the STD is displayed on the PFD before the knob is pushed, the knob selects a preselected barometric setting or the last value that was selected before STD was selected.

- Any barometric reference different from the standard barometric setting is adjusted by rotating the middle BARO knob.

- The outer BARO knob selects inches of mercury or hectopascals as the PFD barometric reference.

Parameter indication controlled on ND

The ND mode selector allows the crew to select one of four map displays (approach, VOR, map, or plan).

The ND CTR (center) switch allows the pilot to display the centered full compass rose for approach, VOR, and map modes.

Two VOR/ADF switches allow the pilot to display VOR or automatic direction finder (ADF) information on the respective ND. In VOR switch position, the ND displays the VOR pointer, frequency, or identification and associated DME information in all modes except plan. When the switch is in ADF position, the ADF pointer and frequency or identification are displayed on the ND in all modes except plan.

Pushbutton switches located at the bottom of the EFIS control panel allow the pilot to select detailed ND information displays:

- WXR (weather radar) displays weather radar information with cloud color coding and range infor-

mation when in the expanded approach or map mode.
- STA (station) displays navigation radio aids.
- WPT (waypoint) displays waypoints in the ND range selector positions in the 10-, 20-, or 40-nmi range.
- APRT (airport) displays airports on all ranges.
- DATA displays the FMC estimated time of arrival, altitude at each waypoint, and altitude constraints at each waypoint.
- POS (position) displays positions of the aircraft calculated by inertial systems and the global positioning system (GPS).
- TERR (terrain) displays terrain data with color coding of elevations.

Several of these groups of information can be displayed on the ND simultaneously.

The EFIS provides the crew with information reflecting current flight conditions and the environment. It is the flight crew's responsibility to be aware of all information displayed and to make timely and correct decisions on the basis of that information.

Crew Errors Resulting from Incorrect Use of Flight Parameter Indication

To safely complete every flight, the crew members of an automated aircraft must perceive, understand, and practically apply all the information presented to them by the electronic flight parameter indication system. But sometimes pilots erroneously decide not to use all available flight-related data. Such negligence can turn a potential emergency into a major disaster.

The following case reflects a crew's unsatisfactory use of the displayed flight information. The aircraft Airbus 320 involved in the accident had electronic equipment and flight parameter indications similar to those described above. The flight crew did not properly use the aircraft's automated displays providing information on adverse weather conditions in and near the airport, and placed the aircraft, themselves, and the passengers in grave disaster.

Case 4: Too late to divert

On September 14, 1993, an A320 aircraft crew performed a scheduled flight from Frankfurt, Germany, to Warsaw, Poland. The cockpit crew consisted of a captain as the pilot flying (PF) in the left seat, who was to be checked, and an instructor pilot in command of the aircraft in the right seat, who checked the PF's activities. The instructor acted as the pilot not flying (PNF).

The captain in the left seat was a 47-year-old male with a total flight time on all airplane types flown (B707, B727, and A320) of 12,778 h and a total flight time on A320 aircraft of 1440 h. This flight was the last stage of a test for this pilot, who was being tested after a 90-day interval on the ground without flying. On the day of this flight, he had spent 5.01 h flying the plane before the accident occurred.

The instructor pilot in the right seat was also a 47-year-old male, with a total flight time on all airplane types flown (B707, B727, and A320) of 11,361 h and a total flight time on A320 aircraft of 1595 h. During the preceding 90 days he had flown for 59.25 h; in the preceding 30 days, 7.67 h; and in the preceding 24 h, 5.01 h.

The final approach of the aircraft commenced at 15:33 UTC (Coordinated Universal Time) while a cold front was passing over the airport area, with cumulonimbus clouds and a heavy rainshower. The crew of another aircraft that landed just before had noticed the front on the weather radar screen. During the approach to land on runway 11, the ATC controller warned the A320 crew that crews of previously landed aircraft had reported windshear conditions in the approach area. According to meteorological reports at the time of landing, the wind was 220° at 10 m/s (meters per second), or 20 knots, variable with a westward tendency, and the wind was gusting at a velocity of 15 m/s (30 knots). The cloud base was about 250 m (750 ft), and the visibility could be decreased to about 3000 to 2500 m because of rainshower. The runway was covered with a 2- to 4-mm layer of water.

Having established visual contact with the ground, the A320 crew members switched off the weather radar that could help to evaluate the situation properly and decided to land without applying the automatic wheel braking system. The crew did not consider the discrepancy between the wind data displayed by the EFIS and the information on the wind given by air traffic services. The crew also did not notice that the tailwind component displayed on the EFIS exceeded the value defined by the aircraft operations manual (AOM) as acceptable for the aircraft.

According to the AOM for that aircraft, with calculated landing weight of 58 tons (metric) and with full landing configuration (flaps and slats fully deployed, gear down), the lowest selectable speed (V_{LS}) was 130 knots. The final approach speed normally exceeds the V_{LS} value by several knots. The airline flight crew operations manual (FCOM) recommended that the flight crew increase the approach speed in windshear conditions by 20 knots, while the Airbus Industries Factory FCOM recommended a speed increase under similar conditions of only 15 knots. At an altitude of about 2800 ft, the aircraft airspeed was 163 knots, while the ground speed was 180 knots. The actual tailwind component at the moment when the aircraft passed over the outer marker (OM) at a radio altitude of about 2100 ft was 25 knots; at touchdown, it was 18 knots, which exceeded the airplane's limitations.

The aircraft passed the middle marker (MM) at an altitude of 278 ft with an airspeed of 147 knots and a ground speed of 168 knots, and continued approach until touchdown with landing configuration (landing gear down, flaps fully deployed to 35°, manual control of thrust and of aircraft flight-control surfaces).

The crew did not consider the possibility that the runway length might not have allowed enough room to suppress the increased kinetic energy of the aircraft caused by the increased approach speed. This miscalculation could have been the result of insufficient description in the airline's flight crew manual of the methods to be used to determine the actual rollout distance when landing with increased speed.

During the final phase of landing flare, the crew continued to maintain a right bank to compensate for a left drift caused by the crosswind. After passing over the threshold, the aircraft appeared to be distinctly above the ILS glide path, because the crew not only maintained increased airspeed due to windshear but also kept the increase in engine thrust down to a relatively low altitude of 6 ft. This caused significant extension of the flare-out phase, and the aircraft made its first contact with the runway by its right landing gear at a significant distance [770 m (2500 ft)] from the runway threshold.

On this A320 aircraft, the ground spoilers could have been extended if at least one of two conditions had been met: either both main landing gear shock absorbers were compressed or the wheel rotation speed was maintained above 72 knots on both main landing gears. Engine reversers could also have been deployed if shock absorbers were compressed at both main landing gears. The wheel brake activation also depended on the wheel rotation speed of 72 knots. Neither of these conditions was met immediately after the aircraft touched down. Because of the A320 aircraft computer logic and without meeting the conditions described above, the crew was unable to activate the aircraft's braking devices.

After the right landing gear first contacted the runway, the pilot flying attempted to use the wheel brakes, but they failed to work. Only when the left landing gear touched the runway did the automatic shock absorber compression system unlock the use of ground spoilers and engine thrust reversers. The spoilers deployed to full angle, the thrust reverser system began to work, but the wheel brakes, which depended on wheel rotation, began to operate after a delay of about 4 s.

Rollout of the aircraft progressed in conditions of heavy rain and with a layer of water on the runway. Hydroplaning, which occurred during the rollout phase, considerably reduced the braking effectiveness. The runway length remaining from the moment when the braking systems began to work was insufficient to enable the aircraft to stop on the runway. Seeing the approaching end of the runway with an obstacle behind it, the pilot flying managed only to swerve the aircraft to the right.

The aircraft rolled over the end of the runway and, after passing 90 m (295 ft), hit the embankment with its left wing; slipped over the embankment, destroying a radio facility aerial located on the embankment; and stopped right behind the embankment. As the result of the collision, the aircraft landing gear and left engine were destroyed. The fuel spilled from broken fuel tanks, caused the aircraft to ignite, killing the instructor pilot and 1 of the 64 passengers.

Accident analysis

A chain of unsafe actions and conditions caused the disaster. The pilots proved to be unprepared for correct assessment of current flight conditions and safe completion of the approach. The airline flight operation managers failed to provide the crew with the ability to perform the flight safely.

Unsafe actions: Routine violations combined with skill-based and decision errors

The unsafe actions of two people—the captain and the instructor pilot—were deemed to be the immediate causes of the accident. These unsafe actions were expressed in violations and errors made by each pilot as an individual professional.

1. The pilot flying (PF) decided to land with a tailwind component considerably exceeding the aircraft operational limit of 10 knots. The aircraft landing was continued in spite of the excessive tailwind component and an excessive ground speed, and the touchdown point moved considerably ahead toward the end of the runway. The pilot in command silently approved this decision. These actions can be qualified as routine violations made by both pilots because the actual wind conditions were not severe but were well above the aircraft limitations.

2. The steering technique applied in the course of aircraft landing in the touchdown phase can be considered a skill-based error. The captain had to know about conditions that allow the aircraft automation to activate the braking devices. He utilized a right bank as a coun-

termeasure to balance the lateral wind component but failed to eliminate the bank just before touchdown. The error resulted in aircraft touchdown on only the right undercarriage leg and prevented the immediate activation of the aircraft braking devices.

3. The pilot performing the check made a chain of decision errors. He allowed the checked pilot, who had not flown independently for 90 days, to perform the approach and landing in marginal flight conditions. He did not help him make decisions at important phases of the approach. He also did not make his own decision to abandon the landing attempt and to perform a go-around procedure that was justified and feasible.

Preconditions for unsafe actions: Poor crew resource management and leadership
Obviously the pilots of the A320 aircraft did not intend to cause the accident. They also were in normal physical and mental condition, which allowed them to perform the flight. But in fact they violated the rules and made mistakes that led to the disaster. The preconditions that led to these unsafe actions were attributed to substandard practice of flight operations resulting from an unsatisfactory level of crew resource management and leadership.

- During approach the crew did not use all available resources to alert it to actual flight conditions. On achieving a visual contact with the ground, the crew switched the weather radar device off and thus did not notice the atmospheric front accompanied by the windshear indicated on the radar screen.
- In conditions of possible poor braking action, the captain decided to land without employing the

automatic wheel braking system, which is especially useful on precipitation-contaminated runways.

- The flight crew did not react to the change of wind direction and velocity displayed on the EFIS or to the difference between airspeed and ground speed during final approach. The EFIS is a powerful source of information about actual wind and aircraft speed values. The crew had to use these displays to control the aircraft flight path parameters and to make important decisions, especially during the final phase of flight.
- The instructor pilot failed to exercise his leadership by allowing the captain to commit violations and make wrong decisions, and by not making his own correct decisions.

Unsafe supervision: Planned flights of inappropriately trained crew

To ensure safety, flights must be properly organized and supervised by flight operations managers. This requirement is especially important for flight operations of automatic aircraft that are controlled by only two pilots. When performing any flights other than training exercises without passengers, both pilots must have not only generally sufficient flight experience but also a good grasp of the current level of professional skills. Alas, this requirement was not followed in preparation of the A320 flight.

Although adverse weather conditions were forecasted over Poland, the flight mission proceeded as planned, performed by a crew that had a temporarily reduced level of proficiency. The captain had not flown for 3 months. For 30 days prior to the day of the flight, the instructor pilot had had only 2.5 h of flight time.

Also, the flight conditions encountered by the crew during its last flight were probably well beyond its current professional abilities, and inadequate flight supervision resulted in unsafe, inappropriate operations planned for a crew with possibly temporarily impaired flying skills.

Organizational influences: Erroneous flight recommendation documentation
Pilots must strictly follow all established company rules, requirements, and recommendations concerning conditions and limitations of flight operations. In the case described above, discrepancies between the company FCOM and the aircraft manufacturer FCOM speed increase recommendations in windshear conditions could have influenced the crew's decision to maintain an extremely high speed during the final approach. The company flight crew manual did not recommend an adequate method of rollout length calculation as a function of the increased speed on final approach.

The lack of documented recommendations left the crew unprepared to calculate the needed runway length in the actual flight conditions and might even have caused the crew to neglect the approximate landing roll calculations. These factors can be considered an organizational failure resulting in less-than-satisfactory organizational process.

Accident could have been prevented
Automated aircraft have a powerful system of flight parameter indication, namely, the electronic flight instrument system (EFIS). This system provides the flight crew with a wide scope of data that reflect actual and forecasted conditions of the aircraft flight. The main parts of the EFIS are the primary flight display (PFD), the navigation display (ND), and the EFIS control panel.

On large commercial aircraft, each pilot has all three devices and independently operates them.

Closer pilot flight supervision and discipline and good crew resource management could have saved the flight. The A320 aircraft EFIS could have supplied both pilots with all the information needed to complete a safe flight. The crew members could have used the ND indications to obtain more realistic information about conditions of their approach to land in Warsaw. The radar image could have shown them the atmospheric front that induced the windshear conditions. Wind direction and wind speed indications would have informed the crew about actual tailwind and its velocity. Ground speed data, combined with airspeed indications, also contained the information that would have cautioned the crew and prompted the captain to make the only correct decision: to employ a go-around maneuver and proceed to an alternate airport or, if the amount of fuel on board permitted, to wait in a holding pattern until the weather improved.

Automated aircraft do not tolerate passive pilot behavior. Both pilots must be fully involved in collecting and processing information and quick, accurate decision making. This is especially important in adverse flight conditions. In the A320 flight, the captain might have been overloaded by the information that had to be correctly perceived and processed. If the instructor pilot had correctly assessed the flight situation, guided the captain, and decided to abandon the landing approach (and thus expressed adequate leadership abilities), the flight could have been saved.

Flight crew mission planning is extremely important in automated aircraft flight operations. Both the left and the right pilot must possess flight skills that meet the current requirements and must be able to satisfactorily perform all flight procedures. Very good full flight simulators are available for all types of automated aircraft, including the A320. The flight could also have been saved if the flight operations managers had exposed the captain to proper simulator training before assigning him to the flight.

And finally, the top managers of the airline could have saved the flight by establishing a policy that would require description of all operational conditions in the company documentation (FCOM) and strict adherence to those documents. In such a case the crew would probably have correctly calculated the actual tailwind speed and the required runway length and would have decided to perform a much safer go-around maneuver, which would have been more appropriate under the prevailing weather and other conditions.

Pilot's Priority List: Procurement and Analysis of Flight-Relevant Information

No two flights are identical. All flights are unique. But there is a common requirement that must be followed in every flight—all available flight-relevant information must be continually collected, carefully assessed, and properly applied by the flight crew. A healthy dose of alert and professional curiosity throughout the entire flight cannot be harmful to any pilot, regardless of that individual's experience and the aircraft's capabilities.

5

Electronic Crew Warning Systems

In addition to readiness for immediate and effective action in response to aircraft system failures, which fortunately have become much less frequent than in the 1950s, 1960s, and so, there is another reason why pilots must be ready to react immediately to save aircraft and human lives. This reason is related to the environment in which flights are performed.

These days three flight safety problems can be called the most critical: collisions of aircraft in good order with the Earth's surface, loss of airplane control in flight primarily as a result of severe turbulence, and midair airplane collisions. Accidents connected with these problems must be prevented by all available means.

Operation of automated aircraft by a crew consisting of only two pilots requires the crew to remain alert, and respond quickly, to flight situations fraught with a potential compromise of flight safety. Modern aircraft are equipped with electronic systems that inform the crew members about potential danger and provide guidance to avoid it.

Ground Proximity Warning System

The *ground proximity warning system* (GPWS) provides the crew with information about potentially dangerous flight conditions that threaten an imminent impact with the ground. The GPWS is one of the most flight-safety-critical systems to be developed since 1970. Although early examples of this system were not as sophisticated as modern ones, GPWS history has proved the usefulness of this system in avoiding controlled-flight-into-terrain (CFIT) accidents, when an entirely operable aircraft collides with the ground for various nonmechanical reasons, resulting mainly from incorrect human actions in flight or on the ground. In 1974 the U.S. Federal Aviation Administration (FAA) mandated the GPWS for airplane certification. The International Civil Aviation Organization (ICAO) recommended that operators install this system on their airplanes (Bresley and Egilsrud 1997). Operation of early GPWS systems was based entirely on receiving airplane radio altimeter signals. Although these systems dramatically reduced the rate of CFIT accidents, they could not sense terrain ahead of the airplane. In case of a quickly rising terrain, the alert signal was too late to allow the flight crew to effectively prevent the collision. To eliminate this problem, an enhanced GPWS in addition to radio-altitude-based alerts offered two additional features: look-ahead terrain alerting and terrain display.

Modern GPWSs provide all three categories of ground proximity alerts: radio-altitude-based alerts, look-ahead terrain alerts, and terrain display.

Radio-altitude-based alerts

Radio-altitude-based alerts are formed as the result of processing the data obtained from radio altimeters, barometric altimeters, airspeed, glideslope, and airplane configuration sensors. This category of alerts includes the following features:

- Excessive descent rate
- Excessive terrain closure rate
- Altitude loss after takeoff or go-around maneuver
- Unsafe terrain clearance when the airplane is not in landing configuration
- Excessive deviation below an ILS glideslope

The alerts are provided to the crew in both aural and visual forms. An aural alert "don't sink" message and a GND PROX light indicate excessive altitude loss after take-off or go-around.

An excessive terrain closure rate triggers the "terrain" aural alert and the GND PROX light. If an excessive terrain closure rate continues and landing gear and/or flaps are not in landing configuration, the "pull up" aural alert and PULL UP message displayed on both PFDs accompanied by master warning lights are provided to the crew.

An excessive descent rate triggers the "sink rate" aural alert and the GND PROX light. The "glideslope" aural alert and the GND PROX light signal deviation below glide-slope. The volume and repetition rate of this warning message increase as the deviation increases. The warning can be canceled or inhibited below 1000 ft radio altimeter height by a special switch.

An unsafe terrain clearance at high airspeed with either landing gear not down or flaps not in landing position triggers the "too low, terrain" aural alert and the GND PROX light.

Unsafe terrain clearances at low airspeed with flaps not in landing position or with gear not down are accompanied by aural alerts "too low, flaps" or "too low, gear" and the GND PROX light.

Look-ahead terrain alerts

Look-ahead terrain alerts are a comparatively new category of crew warning signals that became available as a result of aircraft computerization. To produce these alerts, the GPWS has a terrain database that contains detailed terrain data near major airports, as well as data in lesser detail for areas between those airports. The airplane navigation display (ND) shows the terrain within 2000 ft of barometric altitude. Look-ahead terrain alerts are based on airplane position, barometric altitude, vertical flight path, and ground speed. The GPWS computer that calculates an estimated time to the impact with the ground uses these parameters.

An airplane descent below unsafe radio altitude while too far from any airport in the terrain database triggers the aural alert "too low, terrain" together with the GND PROX light. At a point that corresponds to a time interval of 40 to 60 s from the computer-projected impact with terrain, the aural alert "caution terrain" sounds, an amber TERRAIN message appears on the ND, the GND PROX light illuminates, and the terrain image on the ND becomes solid amber. As soon as the airplane approaches the high-terrain area at a distance corresponding to a time interval of 20 to 30 s before the computer-projected impact, the area shown on the ND in amber becomes solid red, an aural alert "terrain, terrain pull up" is heard, and the master warning lights illuminate. Also, a PULL UP message appears on both PFDs and a red TERRAIN message appears on both NDs.

Pilots must know that the terrain indicated by the GPWS ahead of the airplane might exceed available airplane climb performance. A ground proximity caution or warning does not guarantee terrain clearance.

Look-ahead terrain alerts and radio-altitude-based alerts are prioritized according to the level of hazard and the required flight crew reaction time. When the airplane is in a windshear, the actual windshear warning inhibits these alerts.

Terrain display

The enhanced GPWS terrain display increases flight crew awareness of the terrain around and ahead of the airplane. Operation of this system is based on application of information contained in a database for all airports with paved runways 3500 ft or longer and 95 percent of the world's land surface. Local areas around airports are presented in the database by high-resolution data, and areas between airports are presented by data with a lower resolution. The terrain is shown on the cockpit navigation display in dotted patterns of red, amber, and green. The colors indicate the terrain height relative to the current airplane altitude.

The dotted red pattern indicates potentially dangerous terrain ahead of the airplane 2000 ft or greater above the airplane's altitude. The dotted amber pattern indicates a potential terrain hazard ahead of the airplane with altitudes between 500 ft below and 2000 ft above the airplane. The dotted green pattern indicates terrain 500 ft or more below the airplane, which must be taken into consideration by the flight crew. Two different dot densities for amber and green areas provide the contouring effect of the display. The colors change when the airplane climbs or descends. In high rates of descent, the color bands shift in advance of 15 to 30 s. A dotted magenta pattern shows areas without terrain information in the airplane database.

Terrain avoidance maneuver

A *terrain avoidance maneuver* can represent the last possibility for the flight crew to save peoples' lives and the aircraft. Pilots must learn by heart and correctly perform the terrain avoidance procedure established by their airline. A terrain avoidance procedure recommended by the Boeing Company can illustrate the crew actions needed to avoid a terrain collision (Bresley and Egilsrud 1997). The procedure must be immediately accomplished by recall whenever inadvertent contact with the terrain is imminent. Any of the following conditions is regarded as presenting a potential threat for terrain contact:

- Activation of the PULL UP warning
- Other situations resulting in unacceptable flight toward terrain

Actions of the pilot flying must be as follows:

- Disconnect autopilot.
- Disconnect autothrottle(s).
- Aggressively apply maximum thrust.
- Roll wings level and rotate to an initial pitch attitude of 20°.
- Retract speedbrakes.
- If terrain contact remains a threat, continue rotation up to the pitch limit indicator (if available) or stick shaker or initial buffet.
- Do not change gear or flap configuration until sufficient clearance from terrain is assured.
- Monitor radio altimeter for sustained or increasing terrain clearance.
- When clear of the terrain, slowly decrease pitch attitude and accelerate.

Simultaneously, the pilot not flying must
- Assure maximum thrust.
- Verify that all required actions have been completed and call out any omissions.
- Monitor vertical speed and attitude.
- Call out any trend toward terrain contact.

Notes:

1. Aft control column force increases as the airspeed decreases. In all cases, the pitch attitude that results in intermittent stick shaker or initial buffet is the upper pitch attitude limit. Flight at intermittent stick shaker may be required to obtain positive clearance from the terrain. Smooth, steady control will avoid a pitch attitude overshoot and stall.

2. Do not use flight director commands.

3. "Maximum thrust" means maximum certified thrust. On engines without electronic thrust-limiting capability, overboost or firewalling the thrust levers should be considered only during emergency situations when all other available actions have been taken and terrain contact is imminent.

The GPWS with the capabilities described above, called *enhanced GPWS* (EGPWS), has been used in modern aircraft such as Boeing 777. But research and practical introduction of achieved results continue. For example, the AlliedSignal Aerospace and the Boeing companies are developing further improvements of the EGPWS. At some airports their countries' national rules require using altitudes referenced to airport elevation (QFE) instead of mean sea level (QNH), which is more widely used in the world. Current EGPWS operations at airports with the QFE altitude reference may require disabling the look-ahead alerts and terrain display.

This disabling reduces the system flight-safety-critical crew information capabilities. To overcome the problem, alternatives are studied to provide a suitable altitude source for QFE-based operations. Another improvement will enable the EGPWS to issue alerts for synthetic structures such as buildings, towers, and antennas.

Windshear alerts

Sometimes the wind may suddenly and rapidly change its velocity and direction. This weather phenomenon is called *windshear*. Usually it emerges in the vicinity of a thunderstorm, cumulonimbus clouds, other atmospheric events such as rapidly moving atmospheric fronts, or in areas with a complex terrain surface. The windshear in its most dangerous form for aviation, called *microburst*, can override the airplane's ability to maintain altitude. In the 1990s windshear was the seventh most frequent cause of all commercial jet airplane fatalities.

To reduce the worldwide rate of accidents caused by windshear, the FAA in 1987 published the *Windshear Training Aid*, which supplied flight crews with information on how to identify, avoid, and maximize the possibility of recovering from a windshear encounter (McMillan and Peterson 1997). In 1989 the FAA mandated the adoption of reactive windshear systems (RWSs) for installation on commercial airplanes (McMillan and Peterson 1997). This system uses air data and inertial data to determine whether a windshear exists, and provides flight crews with a way to positively identify a windshear after entering it. The FAA also offered several airlines an exemption from the RWS installation requirement if they pursued a forward-looking, predictive windshear system (PWS). The PWS technology permits weather radar systems to scan the atmosphere ahead of the airplane and detect windshear before the airplane enters the windshear zone (McMillan and Peterson 1997).

On the latest automated aircraft, such as the Boeing 777, windshear alerts are provided by the GPWS and by the weather radar. If in flight below 1500 ft of radio altitude, the GPWS detects an excessive windshear at the current airplane position and emits the "airplane in windshear" aural two-tone siren followed by aural "windshear" alerts; then red "windshear" messages appear on both PFDs, and master warning lights illuminate.

The weather radar provides windshear alerts by detecting disturbed air zones prior to entering a windshear. This weather radar function is able to detect a windshear only in air masses containing some level of moisture or particulate matter. The weather radar automatically scans for windshear in flight after takeoff to 1200 ft of radio altitude.

If a windshear is detected close to and directly ahead of the airplane, the "windshear ahead" aural alert sounds, red "windshear" messages emerge on both PFDs and both NDs, a red windshear symbol is shown on the NDs, and master warning lights illuminate.

If a windshear is detected within 1.5 mi and directly ahead of the airplane during the approach to land, the "go around, windshear ahead" aural alert sounds, there are red WINDSHEAR messages on both PFDs and both NDs, a red WINDSHEAR symbol appears on the ND, and master warning lights illuminate.

The weather radars operated these days on highly automated aircraft provide windshear alerts by detecting disturbed air zones prior to entering a windshear. This weather radar function is able to detect a windshear only in air masses containing some level of moisture or particulate matter. The weather radar detects microbursts and other windshears with similar characteristics, but it does not provide alerts for all types of windshear. Therefore, the flight crew must continue to use traditional windshear avoidance methods.

Researchers continue to search for new methods to expand existing aircraft windshear detection abilities. Clear air turbulence at high altitudes is another problem that threatens flight safety of commercial aircraft. Coherent Technologies Inc., an avionics company, has developed an infrared Doppler radar-based airborne turbulence detection and measurement technology for commercial aircraft, which can allow the pilot to monitor and avoid severe turbulent-air zones (Proctor 1998). The new equipment tests have shown the ability to provide even moderate turbulence warnings for up to 30 s. Coherent Technologies has estimated that a turbulence detector built for commercial airplanes would provide 30 to 60 s of warning at cruise altitudes of 30,000 to 40,000 ft and speeds of Mach ≤ 0.8.

Airborne Collision Avoidance System

Airplane midair collisions create another flight safety problem. This problem was seriously considered as long ago as 1955, when the Air Transport Association of America (ATA) on behalf of its member airlines initiated a midair collision avoidance program (Julian 1985). In those days there was no extensive network of ground radars to provide aircraft surveillance and separation. Powerful air traffic computers able to detect potential airplane separation conflicts also were not yet created. In good visibility, safe separation of aircraft depended almost entirely on pilot vigilance. In poor visibility, traffic controllers attempted to provide safe airplane separation using flight crew estimated position radio reports based on data obtained from inaccurate and unreliable ground navigation aids. On June 30, 1956, two large aircraft collided in broad daylight and good visibility conditions near the Grand Canyon (in northwestern Arizona, USA), killing 128 people.

In the early 1970s, the so-called secondary radars came into use. This type of equipment, together with another device—an airplane transponder—could identify an aircraft, determine its altitude, and help the air traffic controller provide flight crews with more reliable airplane separation information. Nevertheless, tragic midair accidents continued to occur. One such event that caused the world aviation society to look more closely at this problem occurred on September 25, 1978. In good daytime visibility conditions, a Pacific Airlines Boeing 727 airliner, while descending to land in San Diego, collided with a small Cessna 172 airplane. As a result of this accident, 182 persons in both airplanes and 7 people on the ground were killed.

Airborne collision avoidance systems have been under development since 1965. Since 1975, transponder signals emanating from traffic in the vicinity of the aircraft have been used for detecting the close presence of other aircraft. This approach paved the way for development of the current *traffic alert and collision avoidance system* (TCAS), which alerts the flight crew to possible conflicting traffic. TCAS interrogates operating transponders in other airplanes, tracks the other airplanes by analyzing the transponder replies, and predicts the other traffic flight paths and positions. Operation of this system is independent of ground-based air traffic control and provides traffic and resolution advisories, and flight path guidance, together with traffic, resolution, proximate traffic, and other traffic displays of the other airplanes to the flight crew.

Resolution advisory

The *resolution advisory* (RA) is a prediction that another airplane will enter the TCAS conflict airspace within approximately 20 or 30 s. When issuing a RA, the TCAS provides the crew with

- A voice alert
- Vertical guidance on the PFD
- Vertical guidance on the vertical speed indication
- The red TRAFFIC message on the ND

The RA requires the crew to immediately change the airplane flight path following the PFD vertical guidance indication. This maneuver is needed to avoid collision with other aircraft.

Traffic advisory

The *traffic advisory* (TA) is a prediction that another airplane will enter the conflict airspace in 35 to 45 s. This advisory is provided to assist the crew in establishing visual contact with the other airplane. In this case the TCAS provides the crew with the voice (aural) alerts "traffic, traffic" and an amber TRAFFIC message on the ND. On obtaining this advisory, the flight crew must increase visual surveillance outside the cockpit.

Proximate traffic display

Proximate traffic display is neither a RA nor a TA but indicates that another airplane is within 6 mi in horizontal plane and 1200 ft vertically. Other traffic is indicated on the ND by a filled white diamond. If the other airplane is providing altitude data, then a data tag is also displayed that contains information about the relative altitude of the other airplane and its climbing or descending vertical direction.

Other traffic display

Other traffic display is an indication of another airplane that is within the ND display limits but is neither an RA, nor a TA, nor proximate traffic. The TCAS other traffic symbol is a hollow white diamond. It turns into the proximate traffic display automatically when within 6 mi distance. If the other airplane is providing altitude data, then a data tag with its relative altitude and vertical movement information is also displayed.

Crew Errors in Use of Warning Systems

Failure to execute timely and effective measures in response to warning system signals is among the most common of all aircraft crew errors. This phenomenon may be a result of somewhat unreliable indications of the very first warning systems. Since the 1990s the development of avionics has dramatically improved the flight emergency indications. Nevertheless, response time varies; some pilots immediately start an escape maneuver after perceiving the first warning signal. In other cases, precious time that could be spent saving lives is lost because of the pilots' hesitation.

Escape maneuvers can be compromised by improper aircraft configuration as well as insufficiently aggressive pilot control inputs. The reason for these factors may be the result of inadequate pilot training combined with insufficient indication of safety-critical flight parameters.

Warning systems are especially important for flight safety in some regions of the world with less developed communications, navigation, weather, and radar services. Pilot complacency and inadequate reaction to warning signals in such conditions can be really dangerous. For example, from 1979 through 1989, 44 major accidents involving large commercial aircraft occurred in South America. Of these 44 accidents, 34 were attributed to pilot errors. They could have been prevented if pilots had proper situational awareness and made timely and correct decisions and taken appropriate actions. The accident described below (Ladkin 1996) shows that the flight crew's improper execution of the GPWS escape maneuver could be an important contributing factor to the cause of the accident.

Case 5: Improper GPWS escape maneuver as a factor contributing to the accident

On December 20, 1995, a Boeing 757 aircraft crew performed a scheduled flight from Miami International Airport of Miami, Florida (USA) to Alfonso Bonilla Aragon International Airport of Cali, Colombia.

The destination airport was located approximately 7.5 mi north of Cali, at an elevation of 3162 ft above mean sea level (MSL) in a long, narrow valley oriented north to south. Mountains extend up to 14,000 ft MSL to the east and west of the valley. At the time of the accident, the airport control tower operated 24 h per day, controlling departing and arriving traffic on runways 01 and 19. Runway 01 had instrument landing system (ILS) and VOR/DME approaches available. Runway 19 had a VOR/DME approach available. Radar coverage and radar services were not available.

Both crew members were physically fit, properly qualified, and certificated to operate the airplane on this flight. For instance, this was the captain's thirteenth flight to Cali. The first officer, on the contrary, had never flown to the airport. However, he had flown to other destinations in South America as an internationally qualified B757/767 first officer.

The flight departure was delayed in Miami because of traffic congestion. The delay might have interfered with the aircraft cabin crew rest-time regulations.

The first officer was the pilot flying; the captain was the pilot not flying and performed the radio communications. There were no difficulties with aids to navigation or with communications equipment.

The aircraft followed the flight plan route until entering Cali approach airspace. It was nighttime in the area. The final part of the route adjacent to the airport was oriented in an approximately north-south direction. In accordance with the flight plan, the crew initially intended to land on runway 01. The ILS approach to runway 01 had been entered into the airplane's flight management system (FMS).

The last weather report for 20:00 local time at Cali received by the flight crew was as follows: winds 160° at 04 knots, visibility more than 10 km, clouds scattered at 1700 and 10,000 ft, temperature 23°C and dew point 18°C, and altimeter (QNH) 29.98 in Hg.

The Cali approach controller, on the basis of actual weather data indicating a calm surface wind, advised the crew to make approach to land on runway 19. The crew members, after a brief discussion, accepted the proposal in an attempt to minimize the effect of the delay on the flight attendants' rest requirements.

As the crew members were experienced pilots, they nevertheless hurriedly decided to accept the controller's proposal without carefully considering the realistic possibilities of safely completing the VOR-DME approach to runway 19 from the actual aircraft position. The first officer only remarked that the crew had to expedite descent to manage the approach. To increase the rate of descent, he used speedbrakes.

Because of significant time limitations, the crew was not prepared for the straight-in approach before crossing the initial approach fix. Several necessary steps were performed improperly or not performed at all. To successfully complete the approach, the crew had to accomplish several important tasks, such as preparing and reviewing the approach chart, entering data into the FMS, verifying new approach data, and recalculating and maintaining new flight path parameters.

The crew members failed to note many factors that could have helped them correct their erroneous decision; they were unable to adequately review and brief the new approach and to orally agree with each other on the flight path change before executing it through the FMS. The pilots also experienced difficulties in locating fixes that were critical to conducting the approach. While continuing the descent, they did not recognize that the aircraft was turning left to fly for at least one minute a heading that formed an approximately right angle with the reported inbound course to runway 19.

The crew members' decision to make a VOR-DME straight-in approach in a mountainous area at night instead of performing the initially planned, much safer, ILS approach was the first sign of their reduced situational awareness. Once they decided to accept the offer to land on runway 19, they displayed even less situational awareness, with regard to the location of required navigation aids and fixes and the aircraft flight path and its proximity to terrain. The flight crew demonstrated unsatisfactory situational awareness until the last seconds of the flight.

In addition to the faulty decision made by the captain during the unsuccessful approach, several additional factors could have negatively influenced the crew's situational awareness:

- It was difficult to see the terrain in nighttime visual conditions.
- The crew became acclimated to the hazards of flying in mountainous terrain.
- The first officer, performing his first flight to Cali, relaxed his vigilance, relying primarily on the captain's experience in operating near that airport.
- Because Cali, Colombia, was not at high altitude, the Cali airport was not listed as a special-circumstance airport, and no additional training or pilot checking was required to operate into it.
- Although practically all flight crews of the airline had received additional CRM training after a previous accident where the aircraft of another airline collided with a mountain, the crew members could not recall the training in a time of high stress and workload.
- Terrain information was not shown on the electronic horizontal situation indicator (EHSI) or graphically portrayed on the approach chart.

Immediately before the impact, the pilots realized that the airplane was heading toward terrain. Within 2 s of the GPWS warning, the crew began to increase the engine thrust. The aircraft pitch was also increased to a stick shaker activation value. But the speedbrakes were not retracted. At 21:42 eastern standard time (20:42 Cali local time), operating under instrument flight rules (IFRs) during a descent from cruise altitude in visual meteorological conditions (VMCs), the aircraft crashed into mountainous terrain covered with forest at about 8900 ft mean sea level. Its 151 passengers, 2 flight crew members, and 6 cabin crew members died; only 4 passengers survived the accident. The accident site was near the town of Buga, 33 mi northeast of the Cali VOR.

Accident analysis

Probable causes of the accident were stated as inadequate planning and execution of the approach, inadequate use of aircraft automation and navigation devices for maintaining situational awareness, and making decisions to discontinue the approach in the flight crew's ongoing efforts to expedite the approach and landing in order to avoid potential delays.

The accident investigation data showed that if the flight crew had retracted the speedbrakes immediately after initiating the escape maneuver, the airplane may well have cleared the trees at the top of the ridge, and the accident would have been avoided.

The crew members placed themselves in a dangerous situation as the result of their incorrect decision made immediately before the approach. Other factors also could have directly influenced the crew's erroneous decisions and actions.

Unsafe actions: Pilots' wrong decision to make straight-in approach and skill-based error in escape maneuver execution

Both the captain and the first officer incorrectly decided to make a straight-in approach in nighttime (poorly illuminated) and time-restricted conditions, and without preparing the aircraft for the approach. This decision error was the result of inadequate crew situational awareness during descent. It resulted in a chain of other procedural errors that further reduced the crew's situational awareness and prevented it from making the and methods of measuring conditions that trigger warning signals had become more correct. So, these days there is no reason to consider that a warning signal is false. The only logically based pilot reaction to any warning signal must be an immediate and correct escape maneuver.

Part 3
Automated Aircraft in Flight

6

Flight Path Control

6

Flight Path Control

The automated aircraft flight path can be controlled in one of two ways: manually, in much the same manner as any other airplane; and automatically, through special automatic devices, which develop and provide immediate control inputs to aircraft flight controls and engines. Commands for some parts of manual control and all automatic flight path control are generated by a special computer system, usually called the *flight management system* (FMS). The FMS optimizes the flight performance; performs navigation calculations; provides lateral and vertical aircraft guidance through pitch, roll, and engine thrust automatic control; and performs various other important calculations and predictions.

The scope of similar tasks in all automated aircraft is usually performed by a special system. In the latest automated aircraft this system may have a different name and may perform more functions. For example, the Boeing 777 aircraft has a more technologically advanced and more powerful airplane information management system (AIMS) that also performs functions inherent to the FMS.

Manual Control of Aircraft Flight Path

Manual control supplements automated aircraft control during flight. It is provided by the pilot's muscular efforts applied to a control column with a control wheel or a sidestick for aircraft pitch and bank movement control, and to rudder pedals for aircraft yaw movement control.

Automated aircraft are flown manually through powerful mechanically or electronically controlled mechanisms that are included in aircraft control systems between the pilot and aircraft control surfaces. These mechanisms produce the power needed to overcome significant aerodynamic forces induced by aircraft control surfaces in the process of flight control. Automated aircraft in which those powerful mechanisms are controlled by electric signals are called *fly-by-wire aircraft*. For the pilot, manual control of a fly-by-wire aircraft does not differ from manual control of any other airplane.

While controlling an automated aircraft manually, the pilot can use an automatically calculated indication of the flight director bars on the primary flight display (PFD). This indication shows the pilot how the flight controls can be deflected to maintain the required flight path. This method of control is called *semiautomatic* or *director control*. In this case the pilot, whose role in maintaining the flight path is reduced to simple tracking-movement actions, is significantly assisted by automatic flight path parameter calculations.

Automatic Control of Aircraft Flight Path

Immediate deflection of automated aircraft control surfaces in automatically controlled flight is provided through power-amplifying devices in accordance with commands developed by a system of automatic control called *autopilot flight director system* (AFDS). Operation of the AFDS is coordinated with the operation of another automatic system called the *autothrottle*, which controls the aircraft engines. These operations, along with other functions, are commanded and coordinated by the flight management system (FMS).

Autopilot Flight Director and Autothrottle System Operation

The AFDS provides automatic control of the aircraft flight path. The autothrottle system automatically controls aircraft engine thrust.

The AFDS can be operated in one of three different ways:

1. The pilot can control the aircraft manually using *flight director* (FD) *bars* only. The FD bars display needed flight-control commands on the pilots' primary flight display (PFD). In this case the crew members are able to manually control the aircraft flight path by centering the bars. The aircraft follows the flight path computed by the AFDS computers on the basis of flight path parameters set by the crew as well as the parameters obtained automatically from other aircraft systems and ground navigation facilities. Pilots use this semiautomatic, or director, method of maintaining the flight path to manually control the aircraft when fully automatic flight is impossible because of aircraft failure or lack of

needed ground equipment. The director control is also used in both conventional and automated aircraft control in situations where automatic control is either impossible for technical reasons, or undesirable, such as during pilot training in manual landing, where a precise maintaining of flight parameters on final descent is important. Pilots are allowed and even advised to fly the aircraft manually using the flight director at times to maintain the required level of manual aircraft handling skills. When performing a flight director flight, pilots are assisted by computers in routine flight path calculations. Most of their work in maintaining the flight path involves the comparatively easy task of centering the FD bars. This provides the crew with time to assess the whole-flight situation and to make better decisions.

2. The aircraft can be controlled automatically by the autopilot engaged in the *control wheel steering* (CWS) mode, a special method of aircraft automation in which the autopilot maintains the same pitch attitude and bank angle values that were in effect when the autopilot initially engaged. The autopilot CWS mode allows the crew to manually change pitch and bank values through aircraft controls without disconnecting the autopilot. This method of flight path control is intermediate between flight director aircraft control and fully automatic flight control. Its availability on definite aircraft types and models is optional and depends on the airline flight operations policy. A similar flight-control method is used in manual control of aircraft that have flight decks equipped with sidesticks instead of conventional column and wheel controls.

3. The autopilot engaged in a *command* mode can fully control the aircraft in flight. Flight path parameters selected by the crew, as well as those calculated by the

aircraft computers, are maintained automatically. One or more autopilots can be used simultaneously in this autopilot mode. The number of simultaneously operated autopilots depends on aircraft automation philosophy and, to a certain degree, on the pilot's personal decision. The AFDS computers send orders to the pitch, roll, and yaw hydraulic actuators and also send flight-control commands to the flight director bars on the pilots' PFDs. This method allows utilization of all aircraft control automation capabilities and advantages. It dramatically increases flight path maintenance precision, reduces the flight crew's workload, and allows aircraft operations in marginally complex weather conditions. All these factors provide contemporary aviation with higher flight safety levels and better flight operations efficiency.

Autopilot flight director system operational modes

The autopilot flight director system is employed as a complex computer control system within strictly defined and highly formalized procedures called *operational modes*.

The AFDS controls the aircraft flight in horizontal and vertical planes through specific modes of operation. These modes are divided into three classes: longitudinal, lateral, and common modes. *Longitudinal modes* are used for airplane control in the vertical plane. *Lateral modes* are used for airplane control in the horizontal plane. In addition to longitudinal and lateral modes, which provide aircraft control separately within vertical and horizontal planes, there are also *common modes*, which enable the AFDS to control the aircraft within both vertical and horizontal planes simultaneously.

Two of the AFDS operational modes, known as the *basic modes,* are used for computation and execution of other, more complex, modes. The *vertical speed* is the basic longitudinal mode, and the *heading hold* is the basic lateral mode.

Longitudinal modes

Longitudinal AFDS modes include

- Vertical speed, marked in flight operation documentation as V/S
- Altitude hold (ALT)
- Altitude capture (ALT* or ALT star in Airbus aircraft, ALT CAP in Boeing aircraft)
- Flight level change (LVL/CH in Airbus aircraft, FLCH in Boeing aircraft)
- Profile (PROF in Airbus aircraft) or vertical navigation (VNAV in Boeing aircraft)
- Preset function (used, e.g., in aircraft built by Airbus)

In the *vertical speed mode,* the basic longitudinal mode, the AFDS maintains the vertical speed value that the airplane had at the moment when the mode initially engaged. The pilot can set and maintain a vertical speed value by selecting it on a special control panel located on the aircraft cockpit glareshield. This panel, called the *flight-control unit* (FCU) in Airbus aircraft, and *mode-control panel* (MCP) in Boeing aircraft, is used to control all AFDS functions.

In some flight situations the vertical speed mode does not provide aircraft airspeed and altitude protection, because the commanded vertical speed value is the only flight parameter maintained by the AFDS in this mode. During descent or climb, the airspeed may increase above a maximum limit or decrease below a minimum limit. If the vertical speed mode is engaged below an altitude previously set on the MCP, the aircraft will descend as long as the mode is active. Thus, when engaging the vertical speed mode, the crew must continuously monitor all flight path parameters and take maximum precautions to maintain them. Nevertheless, in some flight situations the vertical speed mode is the most appropriate and should be used by the flight crew.

The vertical speed mode is normally used when a required altitude change is relatively small and significant changes in engine thrust are not desirable. For example, it may be convenient to use this mode for small altitude adjustments in calm (nonturbulent), level flight, as passengers may be disturbed by thrust changes, which are inherent in other vertical maneuver modes (FLCH or VNAV).

Another situation in which the vertical speed mode is acceptable and even needed would be an approach to land without automatic flight director control on final phase of descent. Such a situation, for example, may arise during a nonprecision instrument approach without glideslope guidance. In this case the crew would have to establish and maintain a vertical speed of descent that ensures that given altitudes will be reached at strictly defined points. At the same time, all other flight parameters in the aircraft must be stabilized. The vertical speed mode is the only mode that allows the crew to quickly adjust the vertical speed of descent without rapidly changing the aircraft flight path parameters or configuration. For example, while using any other AFDS descent mode (FLCH or VNAV), the vertical speed can be changed only if the aircraft airspeed or configuration also changes.

The *altitude capture mode* acquires the altitude selected on the MCP. When the selected altitude is reached, this mode changes to the altitude hold mode.

The *altitude hold mode* maintains an existing altitude if the vertical speed value is zero when the mode initially engages. When the aircraft develops a vertical speed, the AFDS initially stops the climb or descent and then plateaus, maintaining the resulting altitude. This mode can be used when the crew controls the aircraft speed through the MCP while other airplane flight path parameters are automatically stabilized.

The VNAV *mode* couples the aircraft flight management system (FMS) to its AFDS and autothrottle. In this mode the FMS controls the aircraft's vertical navigation and engine thrust in an optimal way chosen by the crew.

The *flight level change mode* allows the crew to change altitude in a flight that is controlled either by the autopilot or manually using the flight director indications. This mode is used when an optimal, automatically calculated airplane climb or descent in the VNAV mode is not possible or not desirable. For example, this may occur in situations where air traffic is so heavy that the crew is unable to start climb or descent at a definite route point calculated by the FMS, and this maneuver commences earlier or later. In this case the crew engages the flight-level change mode and modifies the aircraft flight path in accordance with crew commands set on the MCP. As soon as altitude capture conditions are met, this mode disengages and is replaced by the altitude capture and then the altitude hold modes.

The *preset function mode*, used in Airbus aircraft, is connected with the aircraft vertical flight path control. It allows preliminary selection of the next speed or Mach value, which becomes a new commanded parameter as soon as certain conditions are met. These conditions include engagement of the flight-level change or the altitude hold mode and activation of the altitude capture mode.

The commands to be executed in longitudinal modes are indicated on the primary flight display (PFD) by the pitch flight director bar.

Lateral modes

Lateral AFDS modes include

- Heading hold (HDG in Airbus, HDG HOLD in Boeing)
- Heading selection (HDG SEL)
- Lateral navigation (NAV in Airbus, LNAV in Boeing)
- VOR/localizer or localizer (VOR/LOC in Airbus, LOC in Boeing)

The *heading hold mode* is the basic lateral (horizontal) mode of the AFDS. It maintains the aircraft heading if the bank angle was less than 5° at the moment when the mode initially engaged. If the bank angle was more than 5°, the AFDS initially brings the aircraft wings to horizontal position and then maintains the heading in effect when the bank angle decreases to 5°.

The *heading selection mode* acquires and maintains the heading selected on the MCP. This mode is so frequently used because it allows easy changes of the airplane heading in automatic flight. The heading selection mode is very convenient during a radar-vectored flight, when the air traffic controller (ATC) provides the aircraft crew with directional commands by specifying heading, altitude, and speed values. This method of air navigation is widely used in dense air traffic areas to provide the needed separation intervals (i.e., to avoid collisions) between aircraft. The ATC is provided with an explicit picture of all aircraft flying in the airport vicinity and can optimize their flight paths.

Nevertheless, when the AFDS is in the heading selection mode, the flight crew must always be in control, continuously checking the aeronautical environment and calculating the aircraft position using all available equipment [VOR, NDB (nondirectional radio beacon), TCAS, GPS, HSI (horizontal situation indicator) map mode, etc.]. This provides the crew with a sufficient situational awareness and allows it to notice and correct potential, although very uncommon, ATC errors in a timely manner.

The *lateral navigation mode* allows coupling of the FMS to the AFDS for horizontal navigation control. This mode is normally used throughout most of the flight, between takeoff and initial approach for landing. The FMS changes the aircraft flight direction in accordance with a previously loaded flight plan. Sometimes the lateral navigation mode is used even during the final approach phase of flight, such as when the crew is performing a nonprecision instrument approach.

When the lateral navigation mode is engaged, aircraft position is calculated by the FMS. Before the engagement of the LNAV mode, pilots enter into the FMS data, obtained from aeronautical documentation and required for the position calculations. The calculations must be systematically updated by signals from various aircraft and ground navigation systems. The best update is provided to the FMS by the global positioning system (GPS). To prevent possible errors resulting from incorrect FMS position calculations caused by insufficient update or an equipment failure, the flight crew members must continually check the aircraft's actual position using "raw data" from all available navigation aids.

In the VOR mode the AFDS captures and maintains a VOR course selected by the crew. The VOR is a common ground radio navigation facility. The VOR equipment allows the pilot to determine the aircraft's direction (radial or course) relative to a VOR station location. This AFDS mode allows the aircraft to automatically follow a given direction to or from the VOR station.

In the *localizer mode,* the AFDS captures and maintains a *localizer beam*—a special radio signal with a narrow radiation diagram emitted by an instrument landing ground radio facility called a *localizer.* This mode is normally used for aircraft directional guidance during the precision-instrument approach to land, from the initial approach phase through rollout after landing.

The commands to be executed in lateral modes are indicated on the PFD by the roll flight director bar.

Common modes
Common AFDS modes are used when all flight parameters must be changed in a coordinated manner. In common modes during the takeoff roll, initial climb, final approach, landing roll, and go-around phases of flight, changes in aircraft engine thrust are also synchronized with variations in the horizontal and vertical parameters of the aircraft motion.

Common AFDS modes include

- Takeoff mode
- Land (Airbus) or approach (Boeing) modes
- Go-around mode

Takeoff (TO) mode, which employs both longitudinal and lateral modes, is available either with only flight directors (FDs) engaged or with both FDs and an autopilot in the control wheel steering (CWS) mode engaged. This common mode is impossible with an autopilot in the command mode. To perform a successful takeoff maneuver in any automated aircraft, the pilot must use the flight controls accurately and effectively to maintain stable takeoff roll direction, smooth liftoff rotation, and initial climb stabilization. During these important time-limited phases of flight, the pilot is in full control and the aircraft automation plays a secondary, supplementary role.

In the *takeoff mode,* the pilot's manipulations are the only source of the aircraft flight controls' movements until the autopilot is engaged by the pilot. Flight-director-bars indications on the PFD must not be taken into consideration in this mode until the flight path has been stabilized in the initial climb phase after liftoff. Before this condition is met, all aircraft motion control actions are made by the pilot and are based on the pilot's skill. Attainment of specific takeoff speeds (V_1, V_R, V_2) is also displayed on the PFD and helps the pilot identify safety-critical moments of flight connected with these speed values.

In the *approach mode,* the AFDS captures and maintains localizer and glideslope beams of the airport instrument landing system (ILS), guides the airplane during the landing flare, and provides further guidance to maintain the airplane on the runway axis after touchdown during the landing roll. This mode is used for precision instrument approaches with both manual and automatic flight control. When the pilot controls the airplane manually, the FD bars indicate needed flight-control manipulations. The airplane is on the flight path if FD bars appear in the center of the PFD. During automatic approach, landing, and rollout, the autopilot controls the airplane's movements in accordance with the AFDS commands, fed to the autopilot and simultaneously indicated by FD bars on the PFD. In this case, after touchdown, the pilot must use (always manually) the engine reverse thrust, disengage the autopilot when the airplane's speed reduces to a taxi speed value, and manually taxi the airplane to vacate the runway.

The *go-around mode* provides longitudinal and lateral aircraft guidance during a go-around maneuver with automatic engagement of the corresponding thrust mode. This mode can be engaged by pressing at least one of two go levers, with FD and autopilot on or off. If, at the moment of initiating a go-around maneuver, the FD and the autopilot were not engaged, after the go-around mode is engaged by the go levers, the FD bars are displayed on the PFD and provide flight path guidance. In this case the pilot must perform the maneuver manually. If the autopilot was engaged before the go-around maneuver was initiated, the maneuver is performed fully automatically after go-around mode is engaged.

Autothrottle operational modes

The autothrottle system also can be operated in one of several modes:

- In the *thrust mode*, where the autothrottle provides the thrust needed to maintain the aircraft vertical speed required by an AFDS longitudinal mode
- In the *thrust reference mode*, which provides thrust within a selected thrust limit
- In the *speed mode*, where the autothrottle maintains a selected speed
- In the *idle mode*, which corresponds to the thrust provided when the engines thrust levers are in idle position
- In the *hold mode*, where the thrust levers are not controlled automatically but can be adjusted or manipulated manually

Automation Control Interface

The *automation control interface* enables pilots to control the aircraft through the autopilot flight director system (AFDS) and the autothrottle system. These systems develop the controlling signals required to automatically maintain flight path direction, altitude, and speed parameters. Flight crew members use special devices to effect the interface between these two automatic control systems.

Autopilot flight director system control

Pilots can control autopilot flight director system (AFDS) operation by manipulating controls located on the AFDS mode control panel or by entering needed flight parameters into the flight management system via one of its control display units.

Mode control panel

The *mode control panel* (MCP) provides direct control of AFDS operation. MCP designs of various aircraft may differ slightly, but are similar in automated aircraft of all types. The MCP is located on the cockpit glareshield and has switches, buttons, and turning knobs to arm, engage, and select the AFDS modes, autothrottle modes, and flight path parameters. Selected flight path parameter values can be observed on the MCP in special windows.

The autopilot can be engaged by the corresponding MCP switch. When it is engaged, this is indicated on the PFD. The autopilot is rendered inoperative by pushing either autopilot disengage button on the pilots' control wheels. Using an autopilot disengagement bar on the MCP is another way to disengage the autopilot.

The flight director bars indication on the left and right PFDs is normally provided when a corresponding MCP switch is in ON position. During flight, when a go lever is pushed, the flight director bars are displayed even if the switch is in OFF position.

Control functions of the flight management system

The central device of the flight management system (FMS) is its *flight management computer* (FMC), which provides calculations of flight parameters optimized in accordance with criteria previously programmed or modified by the flight crew. Usually the FMS has two identical FMCs. Under normal conditions, one FMC accomplishes all flight management tasks while the other FMC monitors its performance. The second computer replaces the first one if a system failure occurs.

Original data for the FMC calculations are flight plan data entered by pilots, airplane systems operation data, and navigation data from the aircraft database. The results of FMC calculations are aircraft pitch, roll, and engine thrust commands needed for flying with use of an optimal flight profile. These commands are sent to the flight director, autopilot, and autothrottle.

Control display unit

The *control display unit* (CDU) is an electronic device that provides an interface between the pilot and the flight management system (FMS). The automated flight deck has two or three control display units mounted on the flight deck central pedestal. Each of these units can be used by the crew to communicate with the FMS by entering alphanumeric data through its keypad and reading FMC calculation results on its CRT or LCD screen.

Normally the left pilot operates the left CDU and the right pilot operates the right CDU. The central CDU (if any) can be operated by both pilots, and it may have some different functions. To avoid confusion between FMCs, each CDU should be operated one at a time. During flight, the pilot not flying enters needed data into the CDU, obtains confirmation from the pilot flying, and then executes the entered modification by pressing a special key. Earlier CDU models do not have the execution function. The data entered in the CDU of an earlier model are immediately fed into the FMC and included in its calculations. This can increase the probability of data entry errors, resulting in undesirable flight path deviations; this sort of error can be avoided by closer pilot attention.

A common error in CDU use occurs when both pilots use the left and the right CDUs simultaneously. This action can have two negative consequences: (1) FMC operation can become unstable for several minutes; and (2) much more dangerously, if both pilots use the CDUs simultaneously, neither of them is controlling the aircraft.

The pilot must keep in mind another error-protecting feature of CDU operation: the closer the aircraft is to the airport, the less time and crew attention must be devoted to the CDU. All FMS preparations should be completed when the aircraft is still on the ground before takeoff or at a level flight above 10,000 ft before landing. The crew may be forced to use the CDU while in immediate airfield proximity, such as in cases where the departure procedure has been modified or the landing runway has been changed by the air traffic control. When CDU input modification is absolutely necessary, both pilots must be twice as careful to maintain flight path parameters and the aircraft configuration. The 10,000-ft altitude can be considered as a border between zones of intensive and forced CDU use.

Autothrottle control

Pilots can control autothrottle system operation through the mode control panel (MCP), through thrust mode select panel, or through one of the control display units

(CDUs). In the latest aircraft types, functions of the thrust mode select panel are also performed through the CDU.

The MCP provides control of autothrottle operational modes selection and the aircraft airspeed control when the vertical profile of flight is not calculated automatically. But when the VNAV mode of the AFDS is engaged, the flight management computer (FMC) calculates the autothrottle modes and target thrust values. In this case pilots can control autothrottle operation by entering into the FMC via the CDU all needed flight path parameter changes.

The thrust mode select panel (or a corresponding page in the CDU) is used to select a reference thrust mode (takeoff, go around, climb, maximum continuous thrust, cruise). This device is also used for preliminary selection of an assumed temperature to obtain a degraded reference thrust or selection of fixed derated thrust values to save engine operation life during takeoff.

Flight mode annunciator

The *flight mode annunciator* is a device that informs pilots about aircraft automation modes that are engaged or ready for engagement. Maintaining situational awareness is the most important task of the flight crew. Many automated aircraft incidents and accidents have occurred because of unsatisfactory crew situational awareness of flight conditions. Modern aircraft design provides pilots with information needed for understanding aircraft automation operation.

All engaged modes and some armed modes of the AFDS and the autothrottle are indicated by the *flight mode annunciator* (FMA), which is an essential part of the interface between crew members and aircraft automation. It is represented in three or four (Boeing) or five (Airbus) cells in the upper areas of both left and right primary flight displays (PFDs). Autothrottle modes are annunciated in the left cell. All other cells are used for AFDS annunciations.

Annunciated modes are shown in the FMA cells in two (sometimes three) lines. The upper line shows engaged modes in green color, while armed modes are white (Boeing) or blue (Airbus) and shown in the lower line(s).

There is one important rule, which, when followed, can help pilots to avoid errors resulting from inadequate aircraft automation control and mode selection. Although selection of a mode is accompanied by an indication on the MCP, after any manipulation with the MCP controls, both pilots must use the FMA to verify engagement of the desirable mode. This rule is also applicable to any new selection made on the MCP; pilots must verify the selection on the corresponding indicator (PFD or ND).

In some flight situations AFDS and autothrottle modes change automatically, without pilot intervention through manipulation of the automation controls. This occurs when previously armed modes become engaged, as in glideslope and localizer capture during ILS approach, or when predetermined flight conditions are met, as on reaching a thrust reduction altitude during takeoff with VNAV mode engaged. In such cases pilots must know when a mode change is expected and which mode will be engaged next. They also must verify every automatic mode change on the FMA.

Errors in Flight Path Control

Modern automated aircraft are equipped with a wide array of instruments that allow the flight crew to perceive needed information about the flight path and to reliably control the airplane. The glass cockpit flight parameter indication and flight-control automation are powerful tools to help the crew stay aware of the flight situation and to take right and timely corrective actions. But sometimes flight accidents do occur, usually caused by unsatisfactory use of automated aircraft flight path indication and control.

Case 6: Erroneous flight path control resulting in disaster over Siberia

On March 24, 1994, in nighttime, an A310 aircraft was on a scheduled flight from Moscow, Russia, to Hong Kong. Because the scheduled flight time was more than 8 h, the flight crew consisted of a captain, a relief captain, and a first officer. The 70 passengers on board were serviced by nine cabin crew members; two children—a son and a daughter—of one of the captains were among the passengers.

After 4 h of flight, the relief captain came on duty. After a short time he allowed his children to enter the cockpit and to sit in the left pilot seat: first the daughter, then the son. The aircraft was at an altitude of 10,100 m (33,100 ft), with a speed of Mach 0.80, and was controlled by the autopilot in the lateral navigation (NAV) mode. The first officer was in the right seat. The relief captain stood behind both seats and, after changing the autopilot mode to heading selection, demonstrated aircraft turns to his son. Then he engaged the NAV mode again.

The son fixed the left control wheel in neutral position for approximately 40 s. This manipulation did not allow the autopilot to exercise full control of the aircraft. As a result of the autopilot override, the aircraft began to bank to the right, and this banking steadily increased. At the same time the autopilot continued to maintain the altitude by increasing the aircraft pitch angle.

The crew initially took no corrective actions to restore a safe flight. With the bank slowly increasing, the aircraft began to buffet, and the bank angle reached approximately 55° with a load factor of about 1.7. Finally the aircraft stalled and lost control. After that the crew members, both in their seats, made desperate attempts to restore normal flight, but failed. Although the pilots tried to overcome the stall, they never moved the control column forward from full-back position. The aircraft crashed in a remote area of Siberia. All people on board died on the spot.

Accident analysis

The investigation team summarized the probable causes of the accident as follows:

- The failure of the flight crew to detect the development of potentially hazardous roll attitude and angle of attack, and to take timely corrective actions to avoid a stall
- The failure of the flight crew to make the proper control inputs to reduce the angle of attack and recover from the stalled condition

In addition to the probable causes, contributing factors were determined on the basis of available information. The relief captain's son was allowed to sit in the captain's seat and manipulate flight controls while the autopilot was engaged.

The left control wheel was manually held near neutral position for a ~40-s period, overriding the lateral mode of the autopilot and causing the bank angle to slowly but continuously increase.

The presence of several visitors in the cockpit, the relaxed atmosphere, and the fact that only one qualified pilot had access to the flight controls delayed detection of the impending stall and inhibited recovery once it occurred.

Unsafe actions: Exceptional violation followed by perceptual and skill-based errors

The relief captain severely violated flight operations regulations by allowing an untrained person (his son) to occupy the left pilot seat in the cockpit and by showing him the aircraft control operation. This exceptional violation caused abnormalities in the autopilot performance, which brought the aircraft into stall conditions.

The crew members initially did not notice the autopilot abnormalities, because of improper perception of engaged aircraft automation modes. The pilots' perceptual error allowed further development of an abnormal situation, which grew into an emergency caused by the aircraft stall.

The pilots' attempts to escape the aircraft stall were not successful because they lacked the skill to correctly execute the escape maneuver. To escape a stall, the control column must be moved forward behind its normally neutral position. The investigation data showed that the column was in full back position practically until the impact.

Preconditions for unsafe actions: Misplaced motivation and crew resource mismanagement

The relief captain created a precondition for the accident because he had misplaced his priorities during flight duty. Instead of organizing the flight crew, carefully monitoring its performance, and managing all available crew resources to safely complete the flight, he shifted his attention to his children, trying to show them the aircraft control in a real flight. This dangerous behavior was a result of his poor professional discipline and irresponsible personality.

The first officer failed to back up the relief captain in his proper performance of flight duties. The first officer did not remind the captain that state aviation authorities strictly prohibited any people other than the crew members to enter the cockpit. He also did not monitor the aircraft flight path parameters and automation mode indication and did not inform the relief captain about the indication deviations. The substandard practice of operations shown by the relief captain and the first officer in their poor crew coordination and communication resulted from crew resource mismanagement.

Unsafe supervision: Insufficient crew training
The flight crew members were not sufficiently trained in crew resource management. They failed to observe flight safety standards and maintain proper crew coordination and communication.

The pilots were also insufficiently trained to perform the stall escape maneuver, evidenced by the fact that their actions during the aircraft stall could not result in restoring a normal flight. These factors qualify the crew flight operations managers' unsafe activity as inadequate supervision.

Organizational influences: Possibility of rule violations, unsatisfactory safety programs
The organizational climate and process also partly caused the accident. The airline's organizational climate did not prevent the relief captain from violating the flight operation rules. It is possible that company policy regarding who entered the cockpit during flight was somewhat lax and thus the captain's children were allowed to visit and distract him from his duties.

All flight personnel training programs in an airline must be directed to ensuring that its pilots are able to provide flight safety under normal and abnormal flight conditions. A crew that is unable to cope with an aircraft stall is a sign of an airline's unsatisfactory flight safety training program.

Accident could have been prevented
The relief captain could have prevented the accident if he had been a disciplined pilot, followed the flight operation rules, and organized crew coordination and performance during flight in accordance with crew resource management principles and flight crew operations manual requirements. He also could have prevented the accident if he had known and correctly performed actions that were needed to escape the aircraft stall.

The first officer could have prevented the accident if he had reminded the captain that no one except crew members was allowed in the cockpit, had properly monitored the flight path and the aircraft automation operation modes, and knew how to escape the aircraft stall and told the relief captain what to do, or had performed these preventive measures correctly himself.

The flight operations supervisors could have prevented the accident if they had properly trained the flight crew members in crew resource management and stall escape maneuvers.

The airline's top managers could have prevented the accident by establishing an organizational climate that made it impossible to violate flight operation rules, and if they had organized the airline's working process to ensure that it had satisfactory flight personnel safety training programs.

Pilot's Priority List: Flight Path Monitoring and Control

Automated aircraft are equipped with all devices needed for monitoring and controlling the flight path.

184 Automated Aircraft in Flight

Their pilots have exclusive possibilities to timely perceive and assess any change in flight path parameters, and to take corrective actions to prevent development of a disastrous situation and to save the flight. But this possibility can be realized only if pilots continually monitor an aircraft's flight path and systems parameters, stay in control of the aircraft, and can correctly use the aircraft controls to make the needed actions in a timely manner.

7

Automated Air Navigation

For many years flight crews gathered information about their aircraft positions from every available source. The sources used were maps compared with ground landmarks and reference points, signals of navigation radio facilities, and positions of celestial bodies defined by astronomical instruments. The compiled information was interpreted in various ways to be used for checking the aircraft's progress along the planned track. The relative value of every piece of gathered information was assessed using the crew members' judgment, and necessary changes in heading, altitude, and engine power setting were made. These changes were not continuous and were made at varying intervals. At appropriate moments, positions were reported to air traffic controllers (Moore and Page 1987).

The high air traffic density in some areas of the world and air transportation economic needs required new methods of aircraft navigation. The results of intensive research of all previously isolated air navigation tasks were evaluated and brought together into a single integrated system. The design and creation of electronic flight management systems made the process of aircraft navigation and flight optimization, from aircraft track maintenance to optimum engine power use, more precise, continuous, and automatic.

Aircraft Navigation Electronic Systems

Aircraft navigation electronic systems calculate the aircraft position and other important flight-path-relevant data. To supply flight management systems of automated aircraft with navigation information throughout the flight independently of time of day, ground radio facilities, monitoring of terrain or areas of flying, inertial reference systems, and global positioning systems are used.

Inertial reference system

The *inertial reference system* (IRS) is a fully independent aircraft navigation aid. It is called a "strapdown" inertial system because it performs all calculations relative to the aircraft airframe. This system measures aircraft accelerations and rotations around the three axes (x, y, z), and at any moment it deduces aircraft movement parameters: current position, acceleration, ground speed, vertical speed, track, and heading. The IRS also provides calculations of wind direction and speed and supplies data for the displays, flight management system, flight controls, engine controls, and other systems.

To reach a desired level of calculation precision and operational reliability, an automated airplane inertial reference system normally has three identical inertial reference units (IRUs). Each unit has a sensor and a computer. The sensor contains three laser gyroscopes and three accelerometers that develop electrical signals proportional to any movement of the IRU. Each IRU is mounted with a fixed orientation to the longitudinal, lateral, and vertical axes of the aircraft.

The crew operates the IRS from a special control panel. To initialize the computations, it is necessary to supply the system with an initial aircraft position and to perform the system alignment, which requires maintaining the aircraft motionless in its parking position for a short period of time (about 10 min). The aircraft's initial position is entered into the IRS through a control display unit (CDU) after its alignment is completed.

The navigation mode is the primary mode of IRS operation. This mode is normally engaged automatically after system alignment, and it is used throughout the flight. Another mode, *attitude IRS*, is a supplementary mode. In case the system alignment is lost during flight, the crew can engage this mode to obtain aircraft attitude information.

Global positioning system

The *global positioning system* (GPS) is another navigation system and is based on the principle of simultaneous measurements of distances to several (normally four) artificial Earth satellites. As a result of computer processing of the measurements, very accurate air navigation data and flight path parameters are calculated and supplied to the flight management system (FMS) and other aircraft systems.

The GPSs mounted in automated airplanes operate independently of the crew and engage as soon as the airplane is electrically powered.

Radio navigation systems

In addition to the inertial reference system and the global positioning system in automated aircraft, as well as in other airplanes, conventional radio navigation systems are used. These systems are used for navigation on air routes, in airport areas, and during approaches to land. They include automatic direction finders, very-high-frequency omnidirectional radio ranges (VORs), distance measuring equipment, and marker signal receivers. Taylor has described these systems in a good book (Taylor 1997). In the present book those systems are described only in the minimal amount of space needed to understand automated aircraft flight operations.

Automatic direction finder

The *automatic direction finder* (ADF) is a low- and medium-frequency radio receiver with two antennas. One of them, an open antenna, has a circle directional pattern while another, so-called loop antenna has a figure 8-shaped directional pattern. A summary signal from a radio station received simultaneously by both antennas reaches its power extremes (maximum or minimum) in only one position of the loop antenna. The loop antenna position corresponding to an extreme signal is automatically defined and shown on a cockpit instrument: an ADF indicator. The ADF indication gives a bearing from the airplane to the radio station. In the glass cockpit this indication can be presented on the navigation display (ND).

Low- and medium-frequency navigation signals used for ADF operations can be easily generated, transmitted, and amplified. This makes the corresponding ground and airplane equipment comparatively cheap. But sometimes these signals can be distorted or blocked by various atmospheric phenomena, such as a thunderstorm. Another deficiency of these signals is their instability in mountainous areas as well as at night and during periods of intense sun radiation, especially in high-latitude areas. All these factors can significantly reduce ADF indication precision. Because of these inherent deficiencies, the ADF can be reliably used only in the vicinity of airports, or when emitted signals are sufficiently strong.

Very-high-frequency omnidirectional radio range
To overcome the low- and medium-frequency signal deficiencies, very-high-frequency (VHF) signals are used in radio navigation and radio communication. These signals provide much more stability in navigation parameter indication in all geographic areas and at any time of day. Although these signals can spread only within the direct visibility range, the high altitude of aircraft flights made the VHF systems very popular in modern aviation.

Airplane VOR equipment consists of a radio receiver able to define direction to an emitting station. The working principle of the VOR is different from that of the ADF. Only one antenna is used, and the radio signal is more complex. But these differences are not important for the flight crew. More important is the fact that VOR indications are reliable and stable within a range limit from several dozens to hundreds of nautical miles. The VOR indicator shows the magnetic bearing, or *radial*, from the ground station to the aircraft.

Often the ground VOR station is combined with another VHF navigation facility: a distance measuring equipment station.

Distance measuring equipment
The airplane *distance measuring equipment* (DME) is a device that receives signals of the ground VOR/DME

station and calculates the distance between the aircraft and the station. In the glass cockpit the DME reading is shown on the primary flight display (PFD) and ND. It is used to identify locations of navigation positions (fixes) on airways and in airport areas.

On some aerodromes the DME equipment is also mounted together with the instrument landing system (ILS) to supply the crew with important distance information during the instrument approach to land.

The VOR/DME equipment is mounted in automated aircraft in two sets. Each set can be tuned automatically by the FMS or manually by the pilot. In normal operation the FMS tunes both VOR and the associated DME for aircraft position radio updates. The ND must be in the map or plan mode to allow FMS tuning of the VOR.

Instrument landing system

The *instrument landing system* (ILS) is a specific radio navigation system used to guide the aircraft during descent and runway alignment on final approach to land. The ILS operation principle is based on simultaneous use of two radio signals that are shaped in forms of vertical and sloping planes. Two transmitters radiate the signals in precisely defined directions. One of the signals, called the *localizer beam,* is radiated in the vertical plane coinciding with the runway axis. Another signal, called the *glideslope beam,* is radiated within the sloping plane that crosses the runway surface near the intended touchdown point and forms an angle with the runway equal to an established final approach descent flight path angle. An imaginary line created by the two perpendicular planes forms a radio flight path.

In addition to the localizer and glideslope transmitters, the ground ILS equipment contains outer, middle, and inner markers, and a ground lights system. Outer and middle markers may be substituted for marker radio beacons. DME may be substituted for the outer marker.

In the airplane two radio receivers (a localizer and a glideslope receiver) recognize the airplane's deviations from the radio flight path and develop corresponding deviation signals, which are indicated to the flight crew. In conventional airplanes an instrument containing display components called *position bars* provides the indications. The bar crossing point shows the desired flight path location relative to the airplane center of mass. In the glass cockpit, any localizer and glideslope deviations are indicated on the PFD by symbols in its right (glideslope) and bottom (localizer) segments. The symbol middle position indicates zero deviation. The ILS indications can also be selected on the ND.

Marker signal receiver

To indicate moments when the aircraft passes over important points, special radio transmitters are mounted at those points. These transmitters are called *marker transmitters,* or simply *markers*. All marker transmitters radiate a narrow radio beam vertically from the ground. At the moment when the aircraft passes over the marker, an airplane marker radio receiver indicates this moment by audio and light signals.

Transponder

The air traffic control service needs information about identification and flight altitude of serviced aircraft. This information is transmitted to the controller by a device called a *transponder,* a radio transmitter that radiates reply signals with needed information when the ground radar antenna beam hits the aircraft. Normally two transponders are mounted in automated aircraft. Only one transponder is used in flight; the other is used as backup. Transponders are controlled from a control panel together with the traffic alert and collision avoidance system (TCAS).

Weather radar
The weather radar provides the flight crew with a colored image of precipitation areas in front of the aircraft. The weather radar can also show turbulence zones when sufficient precipitation is present in those zones. The radar can be used as a navigation aid showing mountains, large cities, and water-land borders. In the glass cockpit the radar picture is presented on the ND in its map display mode and some other modes. When the radar is engaged, the weather picture is superimposed on the ND simultaneously with other navigation data.

Flight-Management-System-Controlled Navigation

Automated aircraft do not have a crew member (navigator) who is responsible for correctly following the flight plan along the entire route. Two pilots using the flight management system (FMS) perform all functions of the aircraft navigation. The FMS includes flight management computers (FMCs) that make all needed calculations, and control display units (CDUs) used for interface between pilots and the FMS.

Calculations and optimization of flight navigation parameters in vertical and horizontal planes are primary FMS tasks. The FMC, while making its computations, factors in the flight path constraints imposed by flight operations rules, air traffic controllers, and the flight crew. These constraints may be expressed in forms of altitude and speed limitations. The aircraft weight, balance, and other limitations are also entered into the FMS through the CDU. All FMC calculations are based on the previously entered flight plan. During flight the FMC continuously updates its predictions to ensure that the aircraft proceeds in accordance with the flight plan. Calculated values of flight path parameters such as aircraft heading, airspeed, and vertical speed are sent to the autopilot flight director system and to the autothrottle. If these systems are engaged, they automatically maintain the flight path parameters in accordance with the FMC calculations. The flight crew members communicate with the FMS via control display units (CDUs). Each CDU has a set of visual information presentation forms displayed on its screen. Normally the crew calls display of these forms, termed *pages,* as appropriate. Pressing a corresponding button on the CDU allows a crew member to call a needed page. Sometimes pages are called on the CDU display automatically, for example, during the aircraft transition from climb to level flight.

Flight management system flight preparation

To prepare the flight management system for every flight, pilots must initialize it. Checking on a corresponding CDU page the validity of information stored in the FMS database must precede this operation. After that, pilots enter identifiers of origin and destination airports, the flight number, and other data relevant to the intended flight on an initialization page. Then the actual aircraft position is entered from a reliable source such as navigation documentation or aircraft GPS position indication.

Flight management system initialization for horizontal navigation

Preparation of the flight management system for horizontal navigation begins with entering on the CDU route page portions of airways and airway crossing route waypoints. Entering a standard instrument departure (SID) is the next step of the FMS preparation. It is entered on the *departure* page. Then all route waypoints as well as directions and distances between them must be verified on the *legs* page, and the total route distance between origin and destination airports must be checked on the *progress* page and compared with the similar distance value published in a corresponding navigation document.

The CDU has a special function that allows the pilot to increase the route and leg pages to a number sufficient for the entire flight. Entering wind direction and velocity data at altitudes along the route is the final step in preparation of the FMS horizontal navigation for the flight.

Flight management system initialization for vertical navigation

To prepare the flight management system for vertical navigation, pilots need to have data about the aircraft actual zero fuel weight, amount of fuel on board, and the aircraft center-of-gravity value. The flight crew members obtain these data after completion of aircraft refueling and loading and enter them into the FMS via CDU performance pages. Before this moment the FMS already "knows" the aircraft limitations as well as speed and altitude constraints obtained in the result of the route horizontal programming and the SID entered. Another parameter needed for aircraft performance calculations is an assumed temperature value that is obtained from special tables based on actual takeoff conditions such as the runway length and its elevation above the sea level, outside air temperature, and the aircraft actual takeoff weight. The assumed temperature is a hypothetical temperature "told" to the FMC to calculate a reduced thrust of engines that provides safe takeoff without unnecessary engine stress. Important takeoff speed values (V_1, V_R, V_2) are calculated by pilots or automatically and are also entered into the FMS via the CDU takeoff page. These speed values are displayed on the PFD during takeoff.

Flight management system operation in flight

Immediately before takeoff all manipulations with the CDU must be completed and stopped. CDU pages used by pilots during takeoff are defined by the airline policy and pilots' tasks performed during takeoff. Normally one of the pilots has a page with vertical navigation data (e.g., the *takeoff* page on Airbus aircraft and the *climb* page on Boeing aircraft), and the other pilot has a page that contains horizontal navigation data (e.g., the *legs* page).

The FMS operation in flight begins as soon as the pilot triggers the aircraft automation flight sequence by pressing the go lever on a thrust lever or pushing the takeoff button on the mode-control panel (MCP). From this moment the FMC predicts all waypoint estimates and calculates control commands for the autopilot flight director system (AFDS) and autothrottle to perform the flight in an optimal manner.

Some FMS models do not allow immediate transition from takeoff mode to fully automatic lateral and vertical navigation functions. These models require pilots using the MCP to initially control the aircraft through the AFDS and the autothrottle. After reaching a defined flight path point, pilots engage fully automatic modes of lateral and vertical navigation on the MCP. Only after this moment do they control the flight through the CDU. On aircraft equipped with other FMS models, automatic lateral and vertical navigation functions can be preselected on the ground before takeoff. In this case these functions are engaged automatically on reaching defined points of the flight path, normally identified by the aircraft altitude values. As soon as pilots engage the autopilot, they control the flight through the CDU.

Engagement of the automatic control modes is indicated to each pilot on the flight mode annunciator (FMA) in the upper area of the PFD.

Automated control of flight path parameters

When in flight, after the lateral navigation (LNAV) and vertical navigation (VNAV) modes of the autopilot flight director system (AFDS) have been engaged, the FMS fully controls the aircraft flight path parameters.

Although the FMS provides the aircraft with automatic navigation, pilots can change vertical and lateral flight path parameters by reprogramming the FMC using the corresponding page of the CDU. For example, entering a new speed or Mach number value into the CDU can change the aircraft airspeed. During climb or descent this manipulation will also change the aircraft vertical speed. Lateral flight path parameters can be changed by entering a new waypoint into the CDU route, or by modifying the route to proceed directly to a waypoint that belongs to the route. After any of similar changes, the FMC makes new predictions and performance calculations.

Another feature that allows pilots to change the aircraft flight path parameters without disengaging the automatic control is a speed intervention function. This function is useful when a temporary speed change is needed, as when the aircraft unexpectedly enters an atmospheric turbulence zone, or to expedite climb, or to maintain a speed faster than programmed in the aircraft database. In all these cases the pilot can disengage the FMS speed control by pressing the speed selector on the MCP, and set a desired speed value. All other FMS control functions will stay engaged. If the speed selector is pressed a second time, the FMS will be restored to the previously programmed speed control.

When the pilot needs to change the aircraft track or altitude to be different from the automatically calculated parameter, the MCP controls such as heading select, flight level change, or vertical speed are used. To restore the automatic flight path control, the pilot must press vertical navigation (VNAV) and lateral navigation (LNAV) switches on the MCP. Any automatic mode change that is indicated on the FMA must be continuously monitored by both pilots.

Special CDU pages provide pilots with the information needed to monitor the flight progress and to obtain data needed for position and weather reports, calculating fuel predictions, and changing the altitude. These pages also enable pilots to make needed flight path parameter changes.

Control display unit navigation pages

To provide the interface between pilots and the FMC during aircraft navigation in flight, the control display unit (CDU) has a set of pages for every phase of flight.

The *climb* page has information that describes the flight path during climb phase of flight. It shows the climbed altitude or flight level, speed restrictions during climb, a speed value that can provide the maximum angle of climb, and a speed that provides the most economical flight. Any reasonable speed value can be selected in the speed line, but in this case the flight will not be economically optimal.

The *legs* page shows the sequence of route waypoints with directions and distances between them. For every waypoint, two programmed flight path parameters are shown: airspeed and altitude. The waypoint shown in the upper line of the CDU display is called the "active" waypoint because the aircraft is flying directly to it. This waypoint usually has color coding. Other waypoint-relevant data, such as an estimated time of flight over a specific waypoint, or forecasted wind direction and velocity as well as air temperature above the waypoint, can be easily obtained by prompted transition from the legs page to other pages that list these parameters.

When the aircraft reaches the intended flight level, the climb page automatically changes to the *cruise* page. On this page, in addition to the flight level and airspeed, the following are shown: the engine thrust parameter (N_1); optimum, maximum, and recommended flight levels; estimated time of arrival at destination; and amount of fuel that will remain at that time. Finally, the cruise page allows the pilot to predict when climb to a higher altitude will be possible.

The *progress* page consists of two parts. The first part shows remaining distances to the active waypoint, to the next waypoint, and to the destination, together with estimated time and fuel remaining for each distance. The second part of the progress page shows actual wind data, the flight path horizontal and vertical errors relative to the flight plan, true airspeed, outside air temperature, and fuel consumption data.

The *descent* page automatically replaces the cruise page as soon as the aircraft starts its descent. It shows an end-of-descent altitude and the waypoint above which the altitude will be captured, an economy descent speed, a speed transition value, and an altitude at which the speed transition becomes active to satisfy the destination airport area speed limitations. Any speed restriction that exists at altitudes higher than the end-of-descent altitude is also shown on the descent page.

The *arrivals* page allows the flight crew selection of a standard terminal arrival route (STAR), together with approach and runway selection. This page shows STARs used for approaches at the destination airport, a list of available approaches to land on defined runways with corresponding equipment, and approach transition waypoints.

Aircraft position calculation update

The FMC calculates actual and predicted aircraft flight path parameters on the basis of available navigation information about the aircraft coordinates. The aircraft-coordinates data are supplied to the FMC by one or more sources of coordinates. To avoid automatic navigation errors, these data must be continuously updated.

The global positioning system (GPS) or the inertial reference system (IRS) can be sources of information on coordinates. The GPS provides the most precise aircraft coordinates; so, if available, it is the main supplier of the coordinates. The IRS in this case is used as backup for the GPS.

If an aircraft is not equipped with the GPS or it cannot be used, the IRS supplies the information on aircraft coordinates to the FMC. The IRS, because of its inherent properties, may accumulate some errors. After many hours of flight, these errors may increase to unacceptable values. Thus the IRS output signals during the entire flight must be corrected by using other available sources of reliable position information. The IRS correction in flight is normally provided by VOR and DME receivers that are automatically tuned by the FMS to appropriate ground radio navigation facilities.

Use of the flight management system for approach

Using the flight management system for approach to land makes the aircraft flight path on this route segment—one of the most complex phases of flight—more precise and reliable. The FMS provides pilots with significant help for approach planning and execution. Several important sides of the flight crew activity exist in any approach.

Approach programming

Approach programming of the FMS is used to properly prepare the aircraft's automatic systems for the approach, landing, and a go-around procedure. Normally during flight before or at the beginning of descent, pilots obtain information about the destination airport's runway in use, recommended landing radio facilities, and current weather in the airport vicinity. This information allows the crew to begin to prepare the FMS for the landing approach. To do this, the pilots have to enter the parameters of the intended approach into the FMS. Every instrument approach is performed in accordance with a standard terminal arrival route (STAR) that contains defined waypoints and flight path constraints. A definite runway is appointed for each approach. Definite radio facilities used for the approach can be appointed by the air traffic control (ATC) or chosen by the flight crew. All this approach-relevant information is published in special aeronautical documents. The airline aircraft maintenance personnel will have previously entered this information into the aircraft FMS database.

The first page to be used for FMS approach preparation is the *arrivals* page, on which the crew must enter the chosen STAR with its transition waypoint, the runway, and the type of instrument approach.

Then the *approach reference* page is used to define the reference airspeed value that must be maintained during final approach with the aircraft landing configuration. This speed is determined by entering the aircraft predicted weight during the final phase of the approach. The weight is calculated by subtracting from the actual aircraft weight the weight of fuel that is left to be burned during the remaining flight time. To have the determined reference speed indicated on the PFD during the approach, the pilot must enter its value in the flap speed line of the page. The air pressure reference (QNH or QFE) used in the destination airport is also chosen on this page.

On the *legs* page, the crew must verify the results of the approach waypoints entering against the chosen STAR and the *runway approach plate*, a navigation diagram with radio facilities and required flight path data. The go-around procedure must also be verified. Speeds and altitude values over each approach waypoint must be checked and, if needed, corrected.

If the CDU has the *radio navigation* page, this page is used to tune the needed radio facilities. If the CDU does not have this page, the radio facilities must be tuned manually. Every time a radio facility is tuned, pilots must identify it using the aircraft audio panel.

Landing minimums to be indicated on the PFD must be set on pilots' EFIS control panels.

Automatic approach navigation

The properly prepared FMS allows aircraft navigation in vertical navigation (VNAV) and lateral navigation (LNAV) automatic modes during the entire approach phase until localizer and glideslope landing modes automatically engage. The VNAV and LNAV modes do not require the pilots to manipulate either the CDU or the MCP.

However, if an altitude higher than the glideslope capture altitude is set on the MCP, the pilots have to follow the FMS prompts on the CDU to reset the next altitude on the MCP to continue descending.

To avoid undesirable deviations from the approach route in case of abrupt FMS failure during the approach, the pilots must use available radio navigation facilities (ADF, VOR, and DME) as the FMS backup. Before the final approach phase, they must verify that all needed radio facilities are properly tuned and identified.

To be prepared for radar-vectored approach, the pilots must reset the heading selection bug on the ND to its direct-to position every time the FMS changes the aircraft heading.

Radar-vectored approach
Radar-vectored approach is used in airport areas with very dense air traffic. The fully automatic approach may be terminated by the air traffic controller (ATC). To optimize arrival and departure aircraft sequencing, the ATC often employs radar vectoring, giving the pilots commands to maintain definite headings and altitudes. After receiving a command to follow a definite heading, the pilot must immediately engage the heading select mode on the MCP. This manipulation disengages the LNAV mode, and further aircraft lateral control must be performed from the MCP. Similarly, the ATC command to change the altitude must be followed by setting a new commanded altitude on the MCP and pushing the FLCH (flight level change) switch.

The CDU *legs* page provides on the ND in the map mode the approach schematic picture. It is useful to have the picture unchanged until the ATC commands the crew to proceed directly to an approach waypoint. In this case the pilots should initially turn the aircraft to the waypoint using the HDG SEL (heading selection) knob and the map picture on the ND, reprogram the FMS for the "direct to" navigation, and then engage the LNAV mode. These actions restore the automatic lateral navigation.

Navigation Errors

Flight safety requirements stipulate that pilots must know and always follow principles of reliable air navigation. Clear understanding of what must be done at any moment in preparation for and performance of the flight and knowing how this is to be done can guarantee the flight crew a safe and efficient flight. Nevertheless, abnormal flight operations of automated aircraft often result from flight crew navigation errors.

Flight management system initialization errors

Reliable operation of the flight management system in flight depends on correct preflight initialization of the FMS by the flight crew. Sometimes pilots may make an error in FMS initialization because they are trying to complete the preparation as soon as possible, or they may not be paying sufficient attention to the process. FMS initialization errors made by pilots during preparation for a flight typically include incorrect waypoint entry, waypoint omission, incorrect weight entry, and incorrect entries of SID (standard instrument departure) constraints.

Incorrect waypoint entry

Incorrect waypoint entry occurs when the FMS database has more than one waypoint with a similar identifier, and the waypoint is entered on the legs page. In this case the waypoint must be entered only after its coordinates have been checked using an appropriate air navigation document.

Waypoint omission

A *waypoint* may be *omitted* in the flight plan entered into the FMS because of a database problem or inattentive entry made by the pilot. This error must be discovered and corrected by carefully checking the route entered on the CDU legs page against the approved flight plan hardcopy. The flight crew members must perform this checking routine every time a new flight plan or its new portion is entered.

Incorrect weight entry

An *incorrect weight value entry* in the performance page leads to incorrect takeoff thrust calculations. This error occurs when an incorrect zero fuel weight value is entered in the correct line. It may also occur when pilots enter a zero fuel weight value in the aircraft takeoff weight line. Careful reading of the needed line name before making the entry can prevent such an error.

Incorrect standard instrument departure constraints entry

Incorrect standard instrument departure (SID) constraints entries can lead to violations of noise abatement procedures and more dangerous violations of a safe altitude. This error may occur when pilots have to modify a value automatically entered with the SID. Checking the new value against the departure clearance obtained from the air traffic control (ATC) will help avoid this error.

All these and similar errors can be avoided by good crew coordination and pilot crosschecking throughout the preflight preparation phase.

Route errors

Flight crew errors sometimes occur during a route flight. These errors can result from pilot carelessness in entering data, controlling automatic devices, following flight operation requirements, monitoring indications, and using navigation equipment.

Incorrect waypoint estimate

Knowing the correct estimated time of waypoints passed is one of the flight crew's basic needs. This information is used to make expeditious aircraft position reports, calculate and predict fuel consumption, and perform a lot of other important tasks. The FMC calculates waypoint estimates on the basis of actual flight data entered by the crew into the CDU. Wind directions and velocity together with air temperature at the flight altitude are the most important parameters for estimated time calculations. Even at the same point above the Earth's surface, there is usually some variation in the wind and temperature parameters at different altitudes. These days, before every departure, pilots are provided with sufficiently precise atmospheric forecasts. In this case the probable source of incorrect FMC time calculations is an incorrect atmospheric data entry made by the flight crew during preflight preparation, or flight at an altitude or along a route other than that entered with the wind and temperature data.

To avoid this error, pilots should enter into the FMS the atmospheric data that correspond to the altitude and route actually flown. A useful recommendation may be to periodically compare, during flight, the forecasted wind and temperature values entered earlier into the CDU with the actual values of these parameters as indicated on the cockpit displays.

Forgetful mode selection

Sometimes pilots in automatically controlled flight are advised by the air traffic control (ATC) to temporarily change the aircraft vertical speed, altitude, or heading.

Changing any of these parameters requires engaging an AFDS mode other than VNAV or LNAV. After the ATC restriction is canceled and pilots are cleared to resume normal navigation, they may inattentively leave the vertical speed (V/S), FLCH, or HDG SEL mode engaged without arming an automatic mode. For example, pilots, to resume normal navigation, turn the aircraft toward the flight plan route using the HDG SEL mode, but do not arm the LNAV mode. In this case, although the aircraft initially approaches the needed track, the automatic engagement of the lateral navigation does not occur. After crossing the track, the aircraft diverts from the route with the heading selected just to intercept the track.

Continuous monitoring of the engaged automation modes can help the flight crew avoid this error. On noticing any FMA mode change, the pilot must call out the change. Another pilot must confirm acknowledgment of the message.

Compromised aircraft stability

The aircraft flight operations manual tells pilots how to perform a flight in the most safe and efficient way. For a given flight, a specific optimal altitude is recommended to ensure maximum possible flight economy while maintaining the required safety precautions. Below this altitude, the flight is more expensive; above it, the aircraft's stability may not satisfy established requirements.

To make the flight more comfortable for passengers and even safer in a turbulent atmosphere, the designers of the flight operations manual recommend that the flight crew maintain a definite value of aircraft speed in turbulent-air zones.

If pilots climb above the optimal altitude, or do not establish the recommended speed before entering the atmospheric turbulence zone, they subject the aircraft and all the people in it to an unjustified risk by compromising the aircraft stability. This error must be avoided by exactly following the flight operations manual recommendations.

Belated weather avoidance
Automated aircraft are equipped with weather radars to help pilots quickly discover and avoid undesirable meteorological phenomena such as atmospheric fronts, thunderstorms, and cumulonimbus clouds. But sometimes the aircraft are subjected to icing, hail, and intensive turbulence caused by these factors. One of the possible reasons for this may be the flight crew's negligence in proper use of the radar and erroneous decisions to keep the aircraft on the same course despite the close vicinity of dangerous weather phenomena.

This error must be avoided by continuous radar monitoring during flight through clouds and timely aircraft maneuvering to prevent the aircraft from entering dangerous atmospheric zones.

Improper inertial reference system update
If the aircraft is not equipped with the GPS or if its operation is unreliable, the *inertial reference system* (IRS) is the main source of navigation reference data. Errors in navigation can be induced by incorrect IRS operation if the IRS position data supplied to the FMC are not properly updated. Normally the update is performed automatically from VOR and DME receivers that are tuned by the FMS. If pilots manually tune these devices to obtain some navigation data, the automatic update terminates.

Pilots must avoid errors of this kind by providing a proper IRS update. To provide continuous IRS output update, the VOR and DME controls on pilots' EFIS control panels throughout the flight must be in the AUTO position. Manual operation of the VOR and DME equipment is recommended only for a short period of time, such as for temporarily following a VOR radial due to the ATC command when in a cruise phase of flight, or to perform a nonprecision instrument approach to land.

Approach errors

Navigation errors made by the flight crew during the approach to land are the most dangerous, because in the final phases of approach, the aircraft may be flown at a low altitude, which can be below the heights of surrounding terrain (mountains) or even below buildings, and also because other aircraft may be close to the aircraft at the same altitude. Navigation errors may be caused by pilots' incorrect actions during the approach preparation and executing. Sometimes subjects other than the flight crew provoke these errors.

Incorrect database information

Normally automated aircraft FMS databases have all the information needed for safe and efficient completion of the flight. Nevertheless, sometimes the database information used by pilots may be incorrect or even absent. This can happen when pilots do not properly check database validity before departure and erroneously use old or invalid data.

There may be discrepancies between the information contained in the aircraft FMS database and information found in published navigation documents. These discrepancies can be the result of different update periods established in the airline for electronic and hardcopy information files. Usually hardcopy files are updated more often.

To avoid errors caused by invalid aeronautical data, pilots are advised to check their aeronautical documentation against similar reference documents before departure. During flight, the information derived from the FMS database must be compared with similar data from the hardcopy documentation. The information with the latest date of issue must be used. ATC assistance can be useful for verifying the validity of aeronautical information.

Erroneous aerodrome pressure setting

The destination aerodrome (airport) pressure setting at a transition flight level is a mandatory safety condition of any flight. But sometimes pilots fail to change or to correctly set the aircraft altimeters setting from the standard value (1013.2 hPa or 29.92 in Hg) to a pressure reference value of the destination airport. Many events may lead to the errors: incorrectly received destination weather information, poor lighting in the cockpit at night, time constraints, and simple forgetfulness. The results of any of these errors can be similarly sad.

Two simple rules can help pilots avoid this error: (1) the pressure setting on all altimeters in the cockpit must be checked and confirmed by both pilots, and (2) the approach checklist must have the "altimeter setting" item, and this item must be checked at the transition flight level.

Improper aircraft configuration

The flight crew members manually control aircraft wing flaps, slats, and speedbrakes as well as landing gears. Normally pilots have some kind of indication of the position of the control devices. More than that, if gear or flaps and slats are not prepared for landing, special signals inform pilots about the condition. But sometimes the alert system may be inoperative, or the information it provides may be insufficient to clearly inform pilots about an abnormal aircraft approach configuration. A known example of this kind of error is an approach with speedbrakes deployed down to a low altitude. This error can significantly reduce the aircraft controllability and create a direct threat to the flight safety.

To avoid this error, pilots must check and confirm that the positions of all gear, flaps, and speedbrakes correspond to the flight phase. To ensure that speedbrakes are retracted in time, the pilot flying must maintain a grip on the speedbrake lever all the time the speedbrakes are in use.

Unsatisfactory crew coordination
The pilots of an automated aircraft must perform their duties in accordance with task allocation procedure throughout the flight. One pilot must fully concentrate on maintaining the required flight path parameters, while the other pilot must monitor the aircraft systems and make the needed manipulations with the CDU and other devices. Satisfaction of this requirement during the approach is the most important factor in safe completion of the flight.

Very seldom is an approach conducted in exactly the same way that the flight crew had planned or for which they had prepared the aircraft. The aeronautical environment around the destination airport may change rapidly and may require additional actions from the crew to complete the approach and landing. For automated aircraft flight crew, these changes almost always require corrections in automation programming, tuning for other radio navigation aids, and changing the aircraft track.

To avoid errors that may lead to violations of safe altitudes and safe intervals, in this situation pilots must always stay in control of the aircraft flight path, know where they are flying, and know where they want to fly.

Communication problems
An airplane pilot is not the only person involved in the flight. During the entire flight the pilot has to communi-

cate with other personnel, such as other crew members, ATC controllers, and sometimes with other aircraft crews. Although the communication is very important for flight operations in general, any approach to land of an automated aircraft absolutely requires a proper quality of communication.

Communication between pilots is important because each of the two pilots performs different tasks. A properly coordinated communication is the only means to provide both pilots with a backup for each other to avoid errors in completing their tasks.

ATC controllers have to provide aircraft pilots with vitally important information, and they have to obtain from pilots information that is also important for making correct controller decisions and issuing adequate commands to all flight crews. The type of communication that provides both communicating sides with full and precise information is the only acceptable aviation communication level.

Four factors may significantly reduce quality of communication: poor radio clarity, language difficulties, use of nonstandard phraseology, and the psychological condition of communication participants. Language barriers pose a significant problem in international flying. The companies and individuals involved must overcome this problem by a coordinated effort to learn English, the most widely internationally accepted language.

Aviation, as much as any other complex human activity, has its own terminology, which must be understood by all people who perform professional aviation duties. To avoid communication errors caused by improper understanding in all professional communications and especially in flight operations, only commonly accepted words, terms, and expressions must be used.

Satisfactory communication is an important source of pilots' and ATC controllers' good situational awareness. The flight crew's and the ATC controller's situational awareness, in turn, is the best shield against navigation errors. Pilots aware of their flight paths relative to terrain and other aircraft will never lose control over the flight situation and will not be surprised or frightened by an imminent disaster. A controller aware of the controlled flight situation will be able to provide the crew with timely advice and commands that can save human lives. Regretfully, sometimes other scenarios occur.

Case 7: Deadly navigation errors on approach

On July 31, 1992, an A310 aircraft crew made a scheduled flight to the Tribhuvan international airport near Kathmandu, Nepal, surrounded by high mountains. The highest mountains are located to the north of the airport.

The flight crew commenced a nonprecision instrument VOR/DME approach, named Sierra approach after its initial approach fix (IAF) name, in instrument weather conditions to land on runway 02. During an initial phase of the approach a flap deployment system failure occurred. The pilots initially requested clearance from the ATC to proceed to their alternate airport, Calcutta, but soon normal flap operation was restored, and the pilots changed their minds and reported to the controller that they planned to land at Kathmandu.

However, at this moment the aircraft position was not conducive to a successful straight-in approach, and the crew informed the ATC about its intention to make another approach and requested a left turn to a waypoint called Romeo located 41 nmi (nautical miles) south-southwest of the airport. The crew considered the Romeo point, as was erroneously published in the crew documentation, to be the IAF. The controller, who did not have radar information, allowed the crew to perform another Sierra approach. The crew did not understand this clearance. The Sierra point is located at 202° radial and 16 m from the Kathmandu VOR station.

The flight crew initiated a right (opposite to the initially requested left) turn from the aircraft heading of 025°, commenced a climb from an altitude of 10,500 ft to flight level 180, and reported to the controller about these actions. Then the controller instructed the crew to report 16 nautical miles (nmi) for the Sierra approach, to maintain 11,500 ft, to proceed to the Romeo waypoint, and to switch for radio contact with another controller, the area control center controller. This controller could not utilize indications, among which was the direction from the airport to the aircraft, because the very-high-frequency direction finder (VHFDF) indicator was mounted only in the airport control tower.

While in the right turn, the crew descended to 11,500 ft and, after completing a 360° turn, continued to proceed to the north on a 025° heading. When the flight was in about 5 mi southwest from the VOR, pilots reported to the area center controller that they were on heading 025, wished to proceed to the Romeo, and had some technical problems. The most probable problem at that moment could have been a temporary FMS operating difficulty caused by the pilots' attempts to enter some information into the FMS through two CDUs simultaneously.

The flight continued to proceed toward the north on the 025° heading. At about 16 mi to the north of the airport, the heading was altered to the left by 005°. After a little more than 1 min later, the ground proximity warning system (GPWS) activated and continued to operate until impact 16 s later. The crew's attempts to avoid the collision were thwarted by a delay, because initially the captain assessed the GPWS warnings as false. In any event, the steep terrain would not allow the crew to avoid the collision even if the escape maneuver had been initiated immediately. All 99 passengers and 14 crew members died.

Accident analysis

The following facts relevant to the accident were discovered in the course of investigation.

1. There were misunderstandings between the flight crew and air traffic controllers:
   - The crew request to turn to the Romeo waypoint to initiate another approach was not understood by the controller because the Sierra point was actually the initial approach fix.
   - The crew did not realize that it was cleared for a new Sierra approach, but understood that it was to continue its present approach.
2. Neither pilots nor controllers were aware of the aircraft position:
   - At some point while making the 360° turn the crew became unaware of where the flight was proceeding. The crew traveled in a north-northeast direction, which was opposite the direction of the Romeo fix. The pilots lost awareness of terrain and of the location of ground navigation aids that were in reality behind the aircraft. They did not realize that the aircraft was flying toward, and not away from, high terrain.
   - The crew's interpretation of the FMS navigation data became a problem near the end of the 360° turn. The navigation information displayed

was confusing to the crew members, and they repeatedly attempted to use the FMS to clarify their understanding of the airplane's position.
- Neither controller knew what the actual position of the aircraft was relative to the airport because no direction information was received from the crew. When requesting the aircraft's position, the controllers asked for distance from the VOR, but not radial information.
- The crew heading report of 025° was probably not heard by the area center controller. He also could not use the VHFDF to determine direction to the aircraft at that moment.

3. Poor crew resource management and unsatisfactory crew coordination caused the loss of the crews situational awareness:
    - The crew did not use available VOR, DME, and NDB indications as sources of information for the aircraft's position identification relative to high mountains.
    - The crew's use of the FMS for navigation was uncoordinated and may have led to confusion regarding the system outputs.
    - Approximately 30 s before the terrain impact, the first officer realized that the aircraft was in a potentially dangerous situation and warned the captain in a mitigated manner. The captain ignored the comment.
    - The airplane impacted the terrain while both the captain and first officer were interacting with the FMS.
4. The pilots' workload was significantly increased during the approach because they encountered

unforeseen problems in communication, unfamiliar aircraft maneuvers, and incorrect information:

- There were communication difficulties between the crew and the ATCs as well as with the other aircraft on approach. The problems were caused by poor radio clarity, language difficulties, and the use of nonstandard phraseology.

- The crew did not receive flight simulator training for the Kathmandu airport, even though this facility was identified in the airline as an airport with special operational considerations. This factor increased the crew's workload when it was confronted with the aborted approach.

- Conditions portrayed in the company en route chart differed from the real navigation situation in the Kathmandu area. For instance, the chart showed the Romeo waypoint as the initial approach fix (IAF). The correct IAF was the Sierra waypoint, located much closer to the airport.

The probable causes of the accident were flight crew mismanagement of the aircraft flight path, ineffective radio communication between the area control center controller and the crew, and ineffective crew coordination in conducting flight navigation duties.

Unsafe actions: Skill-based and decision errors made by pilots and controllers

As they were unable to continue the initially intended straight-in approach after temporary failure of the flaps, the pilots did not understand the controller's clearance and failed to make another Sierra approach. These actions can be qualified as skill-based errors. Trying to use the FMS for locating the Romeo waypoint, which was not required for the second approach, the pilots lost their situational awareness. Unaware of the actual flight situation, the pilots erroneously decided to maintain northbound headings, which caused the aircraft to collide with the mountains.

The flight controllers did not have radar information about the flight and made skill-based errors because they did not request VOR radial indication from the crew to identify the aircraft's position. The area center controller could not utilize the VHFDF while the aircraft was in close proximity to the airport, and thus the plane proceeded toward the high mountains. As they were unaware of the actual flight situation, the controllers made decision errors. They did not communicate with the crew members in a manner that could help them restore their situational awareness.

Preconditions for unsafe actions: Substandard practice of flight operations by the airline

The flight crew members did not use all available radio navigation facilities to maintain their situational awareness. They operated the FMS in a manner that could cause its temporary failure. Coordination between the pilots during flight was unsatisfactory. While both pilots were trying to use the FMS for navigation, neither controlled the aircraft flight path. This crew resource mismanagement was the result of substandard practice of flight operations in the airline.

Unsafe supervision: Insufficient crew training and unsatisfactory crew mission planning

Both flight crew members were improperly trained by flight operations supervisors to perform their professional duties. They were not trained in flight simulation practice for flights to the extremely difficult Kathmandu airport. The pilots were not also trained in crew resource management, crew coordination, or communication with each other and with air traffic controllers. These training deficiencies can be qualified as inadequate supervision.

The crew was assigned to the flight mission without proper professional training. The captain's negative reaction to the first officer's mitigated communication of his concern about the potential danger of a mountain collision could be a sign of the captain's supreme domination in the crew. These factors can be qualified as a planned inappropriate operation, because the crew members, in terms of their mutual working effectiveness, were not suited to the mission or to each other.

Organizational influences: Unsatisfactory organizational process

Before the flight the crew had not received proper training. The company navigation documentation had listed incorrect information concerning the Kathmandu airport IAF name and location. These factors could have caused the flight crew's loss of situational awareness and also increased its workload during the second approach attempt. These factors can be qualified as signs of the unsatisfactory organizational process resulting from unsatisfactory crew training standards and supplying the crew with incorrect navigation documentation.

Accident could have been prevented

Although the pilots encountered a temporary aircraft flap control system failure, they could have prevented the accident and safely completed the flight.

The captain could have prevented the accident if he had understood the controller's clearance to proceed to the Sierra waypoint and to initiate another approach. He should have properly organized crew coordination in controlling the aircraft flight path and using all available navigation aids to maintain the crew's situational awareness. He should have heeded the first officer's message about the possible proximity of high mountains, immediately disconnected the autopilot, and aggressively initiated the aircraft to turn to the south to avoid collision.

The first officer could have prevented the accident by ensuring that the captain was aware of the flight situation and by more persistently informing the captain about his suspicions of dangerous terrain proximity.

The air traffic controllers could have prevented the accident if they used all airport and aircraft means, such as VHFDF and VOR, to increase their own and the crew's awareness of the flight situation.

The flight operations managers could have prevented the accident by properly training the flight crew with a flight simulator and in crew resource management. They could also have prevented the accident by not assigning the crew to a flight mission that was too difficult for both of its members.

The airline top managers could have prevented the accident by establishing an organizational process that would provide pilots with proper training and supply them with correct navigation documentation.

Pilot's Priority List: Continuous and Complex Verification of Aircraft Position

No technology achievement can save the automated aircraft from a disaster without good flight crew coordination and reliable communication. Crew resource management became one of the most important aspects of flight crew activity when automation was introduced in the aviation industry.

222 Automated Aircraft in Flight

Automated aircraft have a powerful flight management system capable of providing reliable and correct navigation in flight. But it cannot complete aircraft navigation tasks by itself. Pilots operate this system from the moment that it is initialized until the aircraft is parked after landing. Only the flight crew's timely and correct control actions during the entire period of the FMS operation can guarantee reliable and safe results.

In addition to new electronic navigation aids such as GPS, IRS, and FMS, pilots of automated aircraft have on board well-known conventional navigation systems such as ADF, VOR, DME, and ILS, together with radars, transponders, and radio communication facilities. Pilots must continually use all of them throughout the flight. Only complex utilization of all available navigation aids can guarantee the crew reliable and safe navigation.

Part 4

Operating Aspects of Aircraft Automation

8

Human Role in Automated Aircraft Flight

The preceding chapters of this book were devoted mainly to automated aircraft architecture and modern cockpit design, as well as to principles of automatic flight control and the design and operation of automatic control devices. Explanation of automated navigation, as one of the most significant features of aircraft automation, took a whole chapter (Chap. 7). All this information was gathered, structured, and presented to provide the reader with a systematic understanding of aircraft automation.

Explanations of the technical aspects of automated flight were accompanied by descriptions of pilots' actions needed for competent operation of the equipment. Examples of flight crew's operational errors and aviation mishap analysis were used to consolidate an important part of the material presented in each chapter. But the emphasis on understanding aircraft automation inevitably implicated the human being as the main acting participant in any human-machine automated system in flight operations. The flight crew members' actions were described in earlier chapters in a simplified manner. These actions are described in greater detail in this chapter.

Task Allocation in Automated Flight

Any complex problem to be solved has to be presented as a composite of more simple problems, which, in their turn, can be imagined as a regulated group of tasks. This approach proved to be productive in aviation as in any industry that operates complex systems. The complex problem of modern automated aircraft flight operations can be analyzed as a hierarchy of tasks that have to be completed at corresponding levels. Because of the interdependence between flight operation tasks, a certain degree of coordination between subjects that complete these tasks is also required.

Task allocation concept

The *task allocation concept* is one of the basic approaches to modern aircraft flight crew professional activity organization.

An automated aircraft cockpit crew normally consists of two pilots, who are busy primarily with the aircraft flight path control process throughout the flight. The pilots' relationships with passengers are limited to the captain's short audio announcements during normal flight, and instructions directed to the cabin crew and passengers in case of emergency.

Each of the two pilots has a strictly defined pattern of responsibilities and actions at every phase of the flight. Each cabin crew member, in turn, has a strictly defined zone of responsibilities and actions in the aircraft passenger cabin in normal and abnormal flight situations.

Indeed, tasks performed by each crew member, a pilot or a flight attendant, in automated aircraft must be carefully defined and allocated. This task allocation is designed to provide timely and correct completion of all crew actions needed for safe and efficient flights.

Crew coordination principle

The *crew coordination principle* reflects a balanced functioning of the aircraft crew. The human-machine system in flight operations is usually mentioned as the crew-aircraft system. During an automated flight all components of the crew-aircraft system operate in an optimal coordination between the system parts. This coordination is designed to provide the most efficient performance of the whole system and to fully use all abilities of all system components.

Flight crew members using complex computer systems control an automated aircraft. The aircraft automation systems operate together in a carefully integrated scheme. A mode change in one system is accompanied by corresponding mode changes in other systems. For example, transition of the autopilot flight director system (AFDS) from the flight level change (FLCH) mode to the vertical navigation (VNAV) mode during aircraft climb or descent inevitably leads to a change in the autothrottle system operation. When the FLCH mode is engaged, a speed value set in the mode control panel (MCP) by the pilot controls the engine thrust. When the VNAV mode is engaged, the flight management computer (FMC) operates the autothrottle by defining another optimal speed value.

The same type of coordination is used in the automated aircraft crew activity design. The crew of a commercial aircraft consists of a cockpit crew and a cabin crew. Operation of the aircraft systems by the cockpit crew results in transporting the aircraft from one airport to another. The cabin crew provides the amount of service needed to make the aircraft passengers comfortable during the flight, and also must perform defined actions directed at preserving passenger health and even saving their lives [e.g., performing CPR (cardiopulmonary resuscitation) or passenger evacuation] in normal flight and in case of aircraft emergency.

Roles of the cockpit crew and the cabin crew have changed and have become more formalized in automated aircraft. In addition to aircraft pilots and flight attendants, another participating subject has to be considered: the aircraft automation system. This fact requires an optimal coordination of actions during flight between the aircraft's pilots and its automatic systems as well as between the cockpit crew and the cabin crew. Coordination of automated aircraft systems operation and automatic devices was discussed in earlier chapters. This chapter is devoted to the coordination aspects of the aircraft crew.

Flight deck task allocation

Allocation of tasks in the flight deck performed by the crew and by the automation system is needed to complete definite actions at strictly defined moments of flight. The process of automated aircraft control during flight can be organized in various ways. At least three levels of automation involvement in the aircraft control can be defined:

1. The pilot can control the flight manually through cockpit flight controls without using automated flight-control devices. In this case the designs of certain aircraft require manual control of engine thrust, while the thrust of other automated aircraft can be controlled manually with the autothrottle engaged.

2. The pilot can control the aircraft using the flight-control automation just for the maintaining of stabilized flight path parameters defined by the pilot. The autothrottle in this case may be disengaged, but normally it is engaged. The heading select and the flight level change modes are examples of this level of automation involvement.

3. Finally, the aircraft can be controlled by its flight automation, which fully calculates and maintains the flight path in accordance with a program entered by the pilot. This is the highest possible level of aircraft automation utilization.

All these methods of controlling the aircraft flight path, when properly used, can provide a safe and efficient flight completion. The pilot has only to decide which level of the aircraft automation use is the most appropriate at a given moment in terms of flight safety and efficiency.

Limitations of automation utilization

Certain conditions can limit pilots' ability to utilize the aircraft automation: phase of flight, aviation technology, and flight environment.

Takeoff and landing are phases of flight that impose limitations on utilization of aircraft automation. Aircraft flight path parameters change the most during these flight phases. Even during normal takeoff or landing the control inputs needed for maintaining a desired aircraft path are naturally changing in a wide band. In case of a failure, these input changes may increase dramatically. An example of such a failure is engine failure during takeoff. This failure requires large and simultaneously well-coordinated control inputs to maintain a safe aircraft motion trajectory. A device capable of automatically compensating for the thrust asymmetry caused by engine failure was designed and mounted on one of the latest automated aircraft: Boeing 777. Although this device is very useful and significantly reduces the pilot's workload in case of engine failure, even this modern aircraft must be manually controlled during takeoff.

During all takeoffs of automated aircraft the pilot immediately controls the aircraft flight controls in the cockpit. Automation is allowed to increase engine thrust only to a level programmed by the pilot and to maintain it until meeting a preprogrammed condition (usually this condition is the aircraft altitude). The pilot performs all other aircraft control tasks needed during takeoff: maintaining takeoff roll direction, compensating for a crosswind, rotation of the aircraft by the elevator to increase its wing angle of attack at a defined speed, and maintaining the initial flight path after aircraft liftoff.

Automatic aircraft landing became routine during the 1990s. It can be characterized by employment of ground and aircraft equipment that helps the aircraft automation control the flight path and the pilots to monitor it. The ground equipment may be a limiting factor for aircraft automatic landing. To be useful, this equipment must have strictly defined parameters, redundancy, and self-test systems. If these requirements are not met at an airport runway, the ground system must not be used, and automatic landings on this runway become impossible.

The flight environment may limit the aircraft automation use as a result of air traffic conditions and weather phenomena. The air traffic controller (ATC) may require the crew to follow a given track that is not programmed in the aircraft FMS. In this case, if the aircraft is flown in fully automatic lateral navigation (LNAV) mode, pilots are forced to degrade the level of automation by engaging another, less autonomous AFDS lateral mode: the heading select mode. If another aircraft approaches the distance of the TCAS (traffic alert and collision avoidance system) resolution advisory, pilots must discontinue the automatically controlled flight and manually change the flight path to avoid collision. If an automated aircraft encounters a thunderstorm or heavy air turbulence during flight, the flight crew must change the flight path parameters to ensure safety. If, during an automatic approach to land, the pilots obtain information about an increasing crosswind on the intended landing runway, they may have to make a manual landing, because at the present state of technology the maximum crosswind limits for manual landing are higher than those for automatic landing.

Automation degradation

Automation degradation, or reduction in the amount of automation authority used to control the flight, is an important method of optimizing task allocation in the cockpit when flight conditions change.

The level of automation utilized during flight not only depends on the limitations imposed by the flight phase, technology, or environment. In some cases the pilot may want to use less automation, or employ partial automation degradation, for flight safety reasons. Automation degradation and even automation system disengagement may be needed if the pilot understands that the automation does not produce the results that are expected from it.

In normal flight pilots can temporarily reduce the amount of automation authority to more correctly follow an ATC clearance. For example, when the aircraft is controlled by the AFDS in the LNAV mode, the ATC controller issues the clearance to proceed directly to a waypoint located at an angle to the present aircraft heading. There are two ways to execute this instruction. One of the pilots can make the needed changes in the FMS programming, obtain confirmation from another pilot, and execute the changes. After these adjustments the aircraft will follow the new route. The consequences of these actions may be quite correct, but they can take some time to be realized. In a dense airspace area it can lead to undesirable reduction of safety intervals between airplanes.

Another way to execute the controller's instruction is to temporarily degrade the AFDS operation from the LNAV mode to the less automated HDG SEL mode by pushing the heading control knob on the MCP, followed by turning the knob to initiate the aircraft turn in the new direction. During the initial moments of the turn the FMS can be reprogrammed and the changes can be checked and executed. After that the LNAV mode can be reengaged. The mode capture will occur momentarily because the actual aircraft heading at this moment will be approximately equal to the new heading calculated by the FMS. Three simple additional actions (pushing and then turning the heading button and pushing the LNAV switch) make execution of the controller's command much easier. In certain conditions it may significantly increase the flight safety.

When, for any reason, the pilot understands that the automation does not perform as it should, or if the pilot does not understand what the automation is doing, his or her first reaction must be simple: reduction of the automation level. Automated aircraft pilots must always remember that their aircraft has other, quite reliable, means of navigation that can be used in a manner similar to that used for conventional aircraft. The automation must never be the only means used by pilots to safely complete the flight. It is very important for automated aircraft pilots to systematically practice using conventional radio navigation systems such as NDB, VOR, and

DME and manually flying the aircraft with or without flight director bars indication. Full flight simulator training sessions can be used for keeping pilots current on basic flying and navigation skills.

Pilot's presence in the aircraft control loop

Any controlled system has an organ that continually provides control inputs to maintain the whole system in a desired condition. In turn, the control organ, or controller, receives feedback information inputs that inform the controller about the results of its control actions. This closed pattern formed by control inputs from the controller to the controlled system and by feedback inputs from the system to the control organ is called the *control loop*. The mandatory condition of the reliable realization of a control process is that the control loop always be closed. This means that the controller is immediately informed about the results of its control inputs, as well as about the general condition of the controlled system.

The cockpit crew is the main controlling organ of the aircraft. Although control inputs produced by the cockpit crew may be different according to how they are realized, all those inputs have only one target: the aircraft flight path. It is not very important whether the pilots move the aircraft control surfaces and engine throttles manually, or whether intermediate automatic devices provide movements of flight-control surfaces in accordance with the pilots' commands. What is really important is that those movements correspond to pilots' actions directed toward maintaining the required aircraft flight path.

In any type of manual or automatic aircraft control, pilots must continually monitor the aircraft systems and flight automation parameters, assess the results of this monitoring by comparing perceived flight-relevant information with a model of the flight created in their brains, and take the necessary actions to eliminate undesired differences between perceived information and the model. In other words, the cockpit crew must always be included in the aircraft control loop.

Cockpit crew task allocation

One goal of the automated aircraft designers was to significantly reduce or even eliminate the cockpit crew's needs to perform routine actions and calculations in flight, which required a lot of time and mental efforts, but which were very simple by nature and could be accomplished by automatic machines. This goal has been successfully reached. The tasks of maintaining the aircraft flight path and maintaining the engine parameters are carefully performed by the autopilot flight director and autothrottle systems. Aircraft navigation and performance calculations are reliably provided by the flight management system.

But the cockpit crew functions that require a genuine human quality called "airmanship" at this historic period of time cannot be entrusted to machines. And more than that, the aircraft automation system itself requires human control and management. These considerations are the main reasons for having at least two pilots in any modern automated flight deck.

The cockpit crew of an automated aircraft has two tasks that must be performed during the entire flight. The primary task is to control the aircraft flight path, which means controlling the aircraft's motion in space and time using direct manipulations of aircraft controls in the cockpit or controlling the aircraft's flight using its automatic devices. The main supplementary task is continuous monitoring of the aircraft automation and systems operation.

The aircraft control task requires the pilot's presence in a control loop, in which controlling actions as well as feedback signals cause immediate reactions of the controlled object (aircraft) or the controller (pilot). This control loop is extremely important for the very existence of the aircraft. At some moments during the flight a small failure in the loop may cause irreparable consequences.

The monitoring task can be described in the form of an information loop in which the aircraft automation and system operation information creates input signals for the executive link represented by the second pilot. The pilot processes the input signals and produces output signals to the first pilot, who controls the aircraft. The first pilot makes the required corrections personally or instructs the second pilot to make the corrections needed for normalizing the situation.

Each of these two tasks requires the pilot's continuous, active presence in both loops: (1) the aircraft flight path control loop and (2) its automation and systems monitoring loop. Thus, at least two pilots must perceive important events occurring in both loops, because at some moments of flight efficient performance provided by one pilot in both loops can be difficult or even impossible because of the vast amount of information and tasks that need to be processed and performed.

To provide simultaneous and full completion of the flight-path-control task and the automation and systems monitoring task, the concept of cockpit task allocation between two pilots—the pilot flying and the pilot not flying, or pilot monitoring—has become standard in the automated aircraft flight crew work infrastructure. The exact number of tasks allocated to each pilot depends on the policy of the particular airline. An exemplary pilot task allocation is shown in this chapter. Before each flight the captain decides who of the two pilots will be the pilot flying and who will be the pilot not flying. If conditions during flight so require, the captain can reverse the previous decision.

Pilot flying tasks

In normal flight the pilot flying (PF) controls all parameters of the aircraft flight path. The PF is responsible for

- Maintaining required values of altitude, speed, heading, and engine thrust
- The aircraft configuration
- The aircraft attitude
- The aircraft position indication on the ND (navigation display) in accordance with the flight plan
- Correct and timely instructions of the pilot not flying concerning aircraft systems operation
- The aircraft checklist utilization requests to the pilot not flying

The PF must tune and identify required radio navigation aids, or ask the pilot not flying to perform this operation. In any case both pilots must be informed about radio navigation aids tuning and identifying results.

Pilot not flying tasks

In normal flight, the pilot not flying (PNF) monitors aircraft automation and its systems. At the same time, the PNF performs other functions required for complete operation of the aircraft. The PNF is responsible for

- Continuous monitoring of the aircraft flight path and its systems
- Informing the PF about flight path parameters deviations and systems operation abnormalities in a timely manner

- Radio communication
- The aircraft position indication on the CDU in accordance with the flight plan
- Filing flight documentation
- Obtaining meteorological information
- Utilizing the aircraft checklist at the request of the pilot flying

Boundary values of the flight path parameters that the PF must announce to the PNF are established by the airline. As an example, these boundaries may be as follows: 50 ft from the required altitude, 5° from the required bank and in any case 30° of bank, any speed below the required value or 5 knots above the required speed value, or vertical speed on final descent of more than 1000 ft/min.

Abnormal aircraft systems operation may be indicated by EICAS or ECAM as well as by other indications in the cockpit. The PNF must announce any abnormal system indication.

Cockpit crew coordination

The aircraft automation system is designed so that only one pilot normally controls automatic systems at a time. The other pilot monitors the aircraft system operation and actions of the partner pilot. This circumstance requires a condition in which all actions in the cockpit are understood and approved by both pilots.

Coordination and reciprocal crosschecking between cockpit crew members is a vitally important flight safety element for two-pilot flight crews. Both pilots must continually monitor the aircraft automation mode indication on their FMAs. If any FMA change occurs, pilots must make certain that the required automation mode is engaged and that its indication corresponds to the mode that is currently engaged. Any mode indication change on the FMA must be announced by the PF and confirmed by the PNF. If for any reason the PF did not announce the mode indication change, the PNF must announce it and request confirmation from the PF.

The PNF performs the FMS preflight preparation under the control of the PF, who supplies the PNF with navigation and flight performance data published in corresponding documents.

During flight the PNF enters needed data in the CDU following the PF's requests. The PF then checks the entered data and confirms execution of the modifications. If a waypoint that does not belong to the previously executed flight plan is entered into the FMS, both pilots must ascertain the waypoint's correct coordinates. After any lateral navigation change execution in the FMS the PNF announces, "LNAV available"; this means that the PF can use the AFDS in the LNAV mode. Vertical aircraft navigation in flight using FLCH and V/S modes on the MCP is performed by the PF, who must immediately inform the PNF.

If the flight is performed without engaging the autopilot, the PNF must arm, engage, and switch the needed AFDS modes on the MCP following the PF's requests. The PNF can engage the autopilot at the PF's request, or the PF can do this, if the flight situation so permits, but must inform the PNF beforehand. The autothrottle in automated aircraft is engaged automatically at the initial moment of takeoff by pressing the go lever or the thrust switch. In flight with the autopilot engaged, the PF performs all manipulations on the MCP. If flight conditions require, the autopilot or autothrottle must be disengaged after a corresponding announcement.

The approach-to-land data should be entered into the FMS well before the start of descent because these data influence the FMS descent calculations. Neither pilot should spend too much time using the CDU for navigation during flight in the airport area at low altitudes. The HDG SEL, FLCH, and ALT HOLD modes of the AFDS are more appropriate in these circumstances. The 10,000-ft altitude may be an agreed border between areas of intensive and moderate use of the CDU.

If the pilots exchange their roles during flight, announcements such as 'You have control" and "I have control" are needed to ensure that one of the pilots controls the aircraft at any given moment.

Cabin crew task allocation

The task allocation concept is also used for flight attendants work organizing. Modern commercial aircraft are able to carry several hundred passengers. Even during normal flight conditions, all of these people should be met at the aircraft entrance, shown to their seats, instructed on how to use the seat and its features, how to behave in normal flight and in case of emergency, and served a snack, a beverage, or a meal. A team of flight attendants performs all these duties.

In aircraft of previous generations the cockpit crew consisted of three and even more aviators. One of them could perform some actions in the passenger cabin such as preflight equipment checks or cabin door operation. The automated aircraft cabin crew that consists of flight attendants must now do those tasks without any assistance from the cockpit crew.

An experienced flight attendant called a *purser* heads the cabin crew. The purser communicates with the cockpit crew and organizes flight attendant activities in normal flight and during emergency. Flight attendant activity during flight is coordinated in a manner similar to that of the pilots' activity. It consists of two kinds of tasks performed: safety tasks and service tasks.

Safety tasks

Safety tasks are primary tasks of flight attendants. Although many people believe that large flight attendant teams are hired by airlines to provide their passengers with maximum comfort and service, the main reason for carrying up to 10 and more flight attendants in every aircraft is simple: to ensure flight safety.

In conventional aircraft a flight crew member sometimes performed important flight safety operations outside the cockpit such as closing the aircraft passenger doors before departure and visual inspection of aircraft flight controls, engines, and other systems during flight from the passenger cabin. In case of emergency, part of the old-fashioned flight crew (i.e., the pilots) could help flight attendants locate the source of fire or smoke in the passenger cabin, supply passengers with oxygen, and evacuate passengers after emergency landing. Modern automated aircraft do not allow pilots to leave the cockpit for a significant period of time to perform these operations. This fact made the cabin crew role more important for flight safety, because flight attendants perform practically all crew actions in the passenger cabin during flight.

The minimum number of flight attendants on board is defined by the aircraft designer and normally should not be fewer than the number of aircraft exit doors. In the flight mission, each flight attendant is assigned to a door, and that door is operated and supervised by the attendant; thus the door is under the flight attendant's control during the entire flight.

The modern aircraft door is a complex assembly. It must be strong enough to maintain a significant air pressure from inside the aircraft at a high altitude. It should be easily and effectively closed and opened. Each door is provided with a slide raft that can be used for passenger evacuation and as a floating raft. After being closed before departure, the doors must be prepared for emergency use. Flight attendants, on the purser's command, perform this preparation by arming special door devices. Before the doors are opened after normal arrival, they must be disarmed to prevent slides deployment. Flight attendants also complete this operation.

Flight attendants are carefully trained in their skills to protect passengers' health and even save their lives in case of emergency. Extremely rare but theoretically probable emergency cases include smoke or fire in the cabin, cabin decompression, emergency landing or ditching, and passenger evacuation.

Before takeoff, although passengers are instructed not to smoke in the lavatories at any time, these aircraft compartments, which contain wastepaper and other flammable items, may be sources of smoke and even aircraft fire emergencies. Smoke or a fire on board is extremely dangerous because of the threat of passenger carbon monoxide poisoning by burning gases. Smoke and fire on board are difficult to extinguish because of the continuous air circulation within the aircraft. The best way to fight the smoke or fire on board is to prevent it. For this purpose each aircraft lavatory is equipped with a sensitive smoke detector that can inform cabin crew about traces of smoke in its compartment. Using the smoke detector indication, the flight attendants must quickly localize and extinguish the source of smoke. In addition, they must routinely check all lavatories for potential smoke sources every 15 min. There are strict rules providing aircraft with all usable in-flight fire extinguishing equipment.

Cabin decompression is another potentially dangerous flight situation, which may be caused by an airframe or engine failure. Aircraft decompression at high altitudes used by modern aircraft is extremely dangerous because of the very limited time, measured by dozens of seconds only, before loss of consciousness occurs at these altitudes. Pilots are continually trained to preserve their own consciousness and to perform rapid aircraft descent in case of aircraft decompression to save passengers' lives. Flight attendants are also trained in how to provide passengers with maximum assistance in this situation. Although every passenger is supplied with an automatically deployed individual oxygen mask, the help of a trained crew member can be critically important for some passengers.

An emergency landing or ditching may be inevitable, for example, in case of an inextinguishable fire. In this case flight attendants must instruct passengers to prepare for a possible rough landing, to take the safest position, and to prepare passengers, aircraft equipment, and themselves for possible evacuation.

Emergency passenger evacuation procedure also requires very good cabin crew training and coordination. The behavior of passengers, as human beings, may be very different in emergencies from what it would be in normal life. Some passengers may be unable to move swiftly, or may be frightened. The cabin crew's task in this situation is not only to correctly use the aircraft emergency equipment but also to soothe and organize passengers for rapid and safe evacuation from the aircraft.

And finally, the most common cabin crew flight safety task should be mentioned. Sometimes the flight is performed in a turbulent-air zone. Pilots are trained to find and use flight routes with the lowest possible turbulence. But sometimes it is impossible to fully avoid the atmospheric turbulence. Practically the only way to guard passengers from becoming frightened and suffering injuries caused by the turbulence would be for the flight attendants to carefully check and assist, in a timely manner, every passenger in use of the aircraft safety belts.

Service tasks
Passenger service is the second most important task of the cabin crew. This is one of the most visible aspects of any airline operation. A major part of the airline's profit depends on the service quality provided by its cabin crews. It is not enough for the cabin crew to show passengers to their seats; demonstrate emergency procedures for them; serve them meals, snacks, and drinks; and otherwise ensure that they are satisfied with the airline's flight. To treat every passenger as a valuable person is the cabin crew's job standard in the best airlines.

Big-Team Collaboration

Automated aircraft can be operated efficiently if all people who are connected with this operation act as a well-organized team.

In early aviation history a small group of people operated an airplane: a pilot, a ground mechanic, and a manager who solved all problems connected with organizing and conducting of flights. All these people knew each other personally as well as they knew their colleagues' responsibilities and immediate professional tasks. This closeness and teamwork helped the first aviators overcome difficult unforeseen problems and helped them develop the art of flying, ranging from an exotic and sometimes dangerous hobby to an interesting and profitable business.

Later, when aviation became an industry, independent groups of people interested in special aspects of airplane operation were allotted specific tasks. Flight operations, airplane maintenance, air traffic control, and customer services became separate divisions of an aviation enterprise. These divisions and specializations had a positive impact on the accumulation of aviation knowledge and made flying more safe and effective. A narrow specialization in each professional group became a common requirement for aviators.

Development of information technology changed the operator's role in human-machine systems. Mostly creative tasks were left to human beings, while plenty of routine operations became computerized. As soon as computers penetrated into airports, they changed the working environment of people involved in the aviation industry. And again the need for more close cooperation between aviators emerged, at this time within a highly developed technological environment. At present, the number of people simultaneously involved in the same air transportation process is significantly greater than at the dawn of aviation. But the need for teamwork is the same. Only now the team is much bigger than before.

Flight safety and financial records of several airlines since the 1990s have shown that to achieve socially acceptable working results of automated aircraft operations, a leading group of aviators called *pilots* need close professional relationships with at least three categories of aviators: flight attendants, air traffic controllers, and aircraft maintenance specialists.

Cockpit crew–cabin crew coordination

Flight attendants of an automated aircraft perform many operations during the flight that have to be carefully allocated within flight phases or must be performed in special conditions of flight. For example, preflight passenger instruction (in the use of flotation devices, etc.) must be completed before takeoff, and flight attendants must always ensure that passenger seatbelts are fastened before the aircraft enters a turbulent-air zone. If an emergency occurs, flight attendants' actions must correspond to actual aircraft condition. In case a passenger or flight attendant notices ice on the wing or another abnormality through an aircraft window, or anything unusual in the plane's general behavior or configuration, the cockpit crew must be immediately informed about it. All these factors require effective communication and coordination between pilots and flight attendants throughout the flight.

Indeed, coordination between cockpit crew and cabin crew is an important aspect of aircraft crew activity. Before departure, the aircraft captain presents a preflight briefing for the whole crew, including pilots and flight attendants. During the briefing the captain informs crew members about the estimated duration of the forthcoming flight; describes safety-relevant factors that can be encountered in flight, such as actual and forecasted weather conditions that may cause air turbulence; and defines how information should be exchanged in flight between pilots, flight attendants, and passengers. The captain may discuss with the crew the latest industry and airline flight safety information as well as optimal crew actions in case of an actual emergency.

On board the aircraft the purser informs the captain about cabin equipment availability for the flight, the number of boarded passengers, and special passenger needs (if any). The captain informs the purser about the expected taxiing time duration and possible traffic delays. Flight attendants close the aircraft doors only on the captain's command.

Before takeoff, the pilots inform the cabin crew members by an agreed-on signal of the need to occupy their seats and be ready for takeoff. During flight the purser periodically enters the cockpit to ascertain the flight conditions and to coordinate the cabin crew passenger service activity with the pilots. Pilots have their meals in turn at their seats at appropriate periods of flight. If a pilot needs to leave the cockpit for a short period, the purser or an assigned flight attendant enters the cockpit to provide visual—and, if needed, verbal—contact with the pilot flying the aircraft.

In case of any aircraft abnormality noticed in the cabin by passengers or flight attendants, the purser or any other flight attendant must immediately inform the pilots about it. The captain also informs the purser about possible deviations from the flight plan if these deviations may require flight attendants' assistance. If a significant failure occurs, the captain informs the purser about planned further actions and outlines the whole crew activity in case of possible emergency.

During the final phase of the approach to land, the pilots inform the cabin crew members in a manner similar to that before takeoff about the need to occupy their seats and be ready for landing.

Pilot–air traffic controller interaction

Automated aircraft cockpit crew actively interacts with another group of aviators during flight: the air traffic controllers. Air traffic control (ATC) computerization has significantly changed ATC system operation in developed countries. Powerful computers assist ATC controllers to optimize and assign more direct routes to controlled aircraft. The volume of needed radio communications between aircraft crews and controllers as well as the number of mandatory position reports has also dramatically reduced.

But the growing density of air traffic has brought a negative aspect to this improvement. In some major airport areas the controllers are busy with multiple aircraft communications practically without short time breaks. If a pilot did not understand a controller's message, as originally transmitted, it may be difficult to establish a new contact with the controller to acknowledge the message, because the controller can be continually engaged in communications with other airplanes.

To provide safe and reliable interaction between aircraft crews and ATCs, pilots and controllers must have a thorough, clear understanding of established, common pilot-controller phraseology, concise and understandable reports, and correct and distinct pronunciation. This requirement is especially important.

The pilot-controller radio communication is a valuable resource of situational awareness for both aviators. Possible misunderstandings between them may be caused by language barriers, such as an unusual accent or limited knowledge of spoken English, especially in unusual situations. To avoid navigation errors, both communicating sides (pilots and ATC) must clarify any misunderstanding as soon as possible. Pilots and ATC controllers whose native language is not English must recognize the possible sources of communication errors and improve their knowledge of the English language throughout their professional lives.

Flight crew–maintenance personnel relationships

In addition to cabin crew members and air traffic controllers, the pilots interact with people who prepare the aircraft for their flight. Automated aircraft are fully prepared for each flight by a team of highly qualified maintenance specialists. Not only the airframe and engines, but also all computerized systems of the aircraft control and navigation are carefully checked and served on the ground. To have the maximum information about the aircraft's technical condition and to ensure its safe flight operations, the aircraft crew members and especially the captain must actively collaborate with the maintenance team before as well as after the flight.

Before departure, pilots must perform an aircraft visual inspection and gather information about the aircraft's availability for the flight. This information can be obtained from the aircraft maintenance logbook, a maintenance bulletin, a document called the *minimum equipment list* (MEL), and aural preflight crew briefing made by the maintenance person responsible for releasing the aircraft for the flight.

In the logbook pilots can read messages entered by the previous crew about failures that they encountered. Normally the maintenance team must rectify every failure, and a corresponding message must be provided in the logbook. The maintenance team manager must also release the aircraft for the flight by signing the logbook.

If any failure has not been corrected, a special bulletin informs pilots about the aircraft's current status. A high degree of automated aircraft equipment redundancy and reliability sometimes allows flights to be performed even if the aircraft has failures, as long as these failures are insignificant and are not considered as a potential threat to flight safety. Normally the aircraft can be operated for several days or several flights with these failures. After that allotted time or number of flights, the aircraft must no longer have the failures.

If any failure has not been eliminated, but the aircraft is released for flight, the pilots using the MEL must check the possibility of making the flight. This document has sections for every aircraft system. Each section contains a list of temporary allowed system failures and flight conditions, in which a specific failure can be acceptable. Pilots must determine whether the planned flight satisfies the operational conditions required by the MEL.

The cabin crew also performs the aircraft cabin preflight check, and the purser reports to the captain about the check results and the readiness of the cabin crew and equipment for the flight.

After visual inspection, reading the document inputs, obtaining the purser's report, and listening to the maintenance briefing, the captain, by signing the logbook, certifies acceptance of the airplane for the flight.

Having completed the flight, the pilots must note in the maintenance logbook any aircraft or system failure (including the passenger cabin) or operational abnormalities that occurred during the flight. This will help the maintenance specialists keep the aircraft operable and will provide the next crew with authentic information about the aircraft's status. In some cases the maintenance team may ask for more detailed explanations from the flight crew. The explanations have to be provided in an acceptable form.

Crew Coordination Errors

Modern automated aircraft flight safety requires good coordination between all members of crews. The cockpit crew of only two pilots may be unable to obtain all information needed for safe completion of the flight. In case of emergency, the cabin flight attendant crew can significantly help pilots cope with the failure, make timely and correct decisions, and safely bring their passengers, the crew, and the aircraft to the destination or alternate airport. Alas, people sometimes fail to properly collaborate, and another course of events occurs.

Case 8: Unsatisfactory coordination between pilots and flight attendants as a factor contributing to the accident

At about 8 o'clock in the evening (20:00 hours) on January 8, 1989, a Boeing 737-400 aircraft with 8 crew members and 118 passengers (including 1 infant) on board left London Heathrow airport for Belfast, Northern Ireland.

As the aircraft was climbing through 28,300 ft, an aural warning sounded, as the left-engine compressor began to stall as a result of a fan-blade failure. The autopilot, which had been engaged before, was disconnected. The engine failure resulted in airframe shuddering, ingress of smoke and fumes to the cabin and cockpit, and fluctuation of the engine parameters.

The first officer (PNF) initially commented that there was an engine fire, and after the captain's (PF's) question he answered: "It's the le... it's the right one" (Beaty 1992). The captain ordered the first officer to put the right-engine throttle lever to idle position.

Although the pilots coped well with the stress induced by the failure, they reacted to the emerging engine problem prematurely and in a way that was contrary to their training. They did not assimilate the indications on the engine instrument display before they throttled back the right engine. The shuddering caused by the surging of the left engine ceased as soon as the right engine was throttled back, which persuaded the crew members that they had dealt correctly with the emergency.

The first officer began to read the engine failure and shutdown checklist but was interrupted by air traffic control (ATC) calls and by the commander's own calls to the operating company during which the decision was made to divert to East Midlands airport. The London ATC was then contacted by the first officer, advising them of an emergency, after which the captain asked for the right engine to be shut down. Unnecessary radio calls shattered the flight crew's concentration on the failure and interrupted their necessary actions and dialog sequences.

The first officer then recommenced the checklist, shut down the right engine, and then started the APU. The pilots reprogrammed the flight management system (FMS) for an East Midlands diversion, with which they had some difficulty. The left engine operated apparently normally after the initial period of severe vibration and during the subsequent descent.

After the aircraft started the descent to East Midlands airport, the captain made his first announcement to the passengers, during which he mentioned that they had a problem with their right engine, which had produced some smoke in the cabin. The cabin crew and passengers saw sparks and flames coming from the left engine. The passengers were alarmed by the discrepancy between the captain's announcement and the actual conditions that they were observing, but no member of the cabin crew informed the pilots about it.

The pilots initiated a diversion to East Midlands and received radar direction from ATC to position the aircraft for an instrument approach to land on runway 27. The flight proceeded until the aircraft was on final approach with the landing checklist completed. The approach continued normally, although with a high level of vibration from the left engine, until an abrupt reduction of thrust, followed by a fire warning, occurred on this engine at a point 2.4 nautical miles (nmi) from the runway. It was at this point that the pilots came to understand that they may have stopped the good engine. But their efforts to restart the right engine were not successful.

At 10 s before impact, the captain made his announcement to the passengers to prepare for a crash landing.

The aircraft initially struck a field adjacent to the eastern embankment of the M1 motorway and then suffered a severe impact on the sloping western embankment of the motorway.

In this incident, 39 passengers died in the accident and 8 more passengers died later from their injuries. Of the other 79 occupants, 74 suffered serious injury.

Accident analysis

Investigation of the cause of the accident revealed that the operating crew had shut down the right engine after a fan blade had fractured in the left engine. This engine subsequently suffered a major thrust loss due to secondary fan damage after power had been increased during the final approach to land.

Although the immediate cause of the accident was assumed to be the pilot's decision to shut down the good engine instead of the failed one, the pilots' erroneous actions could have been corrected if somebody from the cabin crew had informed them that they had made the rough mistake.

The combination of heavy engine vibration, noise, airframe shuddering, and an associated smell of fire were outside the flight crew's training and experience. The training of the pilots met the company requirements. However, no flight simulator training had been given, or had been required, on recognition of engine failure on the electronic engine instrument system or on decision-making techniques in the event of failures not covered by standard procedures.

Unsafe actions: Perceptual, decision, and skill-based errors made by pilots and flight attendants

The first officer erroneously informed the captain about the right-engine failure as a result of perceptual error.

This error occurred because the pilots were not trained on a flight simulator to deal with this aircraft modification and to recognize and cope with engine failure.

The captain decided to render idle and then shut down the good right engine without carefully examining the situation. He erroneously informed passengers and flight attendants about the right-engine failure. The captain's decision error and the incorrect information were the results of his unsatisfactory situational awareness.

The flight attendants did not inform the pilots about their erroneous detection of the failed engine. The skill-based error made by the entire cabin crew was a result of the flight attendants' unsatisfactory training in crew resource management in flight.

Preconditions for unsafe actions: Unsatisfactory professional readiness and crew resource mismanagement

Engine failure was new for both pilots. They had never encountered it in flight or in a flight simulator. Thus, they were not professionally ready to rapidly and correctly recognize and cope with the emergency.

The pilots acted hurriedly in the emergency and did not follow operation manual recommendations. While localizing the failure, they were distracted by unnecessary radio communications. The flight attendants did not act together with the pilots as a single team. These preconditions for unsafe actions resulted from crew resource mismanagement caused by substandard practice of flight operations by the airline.

Unsafe supervision: Inadequate crew training

The pilots did not have flight simulator training on how to handle engine failure and other emergency situations. Neither the pilots nor the flight attendants were trained in crew resource management. The crew members' managers had not provided them with sufficient training to guarantee correct handling of emergency situations. The crew's inadequate training can be qualified as a sign of inadequate supervision.

Organizational influences: Company resource mismanagement and inadequate organizational climate

Origins of erroneous actions of pilots and flight attendants as well as causes of inadequate supervision performed by flight operations managers may be found in the company policy. The company did not provide sufficient resources for pilots' and flight attendants' adequate training; this was the result of company resource mismanagement.

The company did not require the necessary cockpit and cabin crew training from the responsible managers. Cockpit crew members and cabin crew members did not act as a united team during the flight. These shortcomings can be considered as results of inadequate organizational climate in the airline.

Accident could have been prevented

The immediate line of defense against this accident was within the cockpit. If the pilots had correctly recognized the failure, they probably would have acted in a manner that would have provided them with a good and operable right engine. This would have enabled the pilots to make a safe landing at the alternate airport.

The second barrier that could have prevented the accident was in the passenger cabin. If any of the 100-plus people in that cabin had been sufficiently insistent to inform the pilots about their mistake after the captain's first announcement, the pilots would have had sufficient time to start up the wrongly shut-down right engine. Of course, the flight attendants had to do that as soon as possible. This would have radically improved the situation.

The pilots' and flight attendants' flight operation managers could have created conditions to prevent the accident. Adequate pilot training in recognizing and coping with engine failure would have allowed them to act correctly and to maintain control of the aircraft until it landed safely. Training of pilots and flight attendants in crew resource management would have changed their responses in a dangerous situation and would have made their actions timely, correct, and well coordinated.

The top management of the airline could also have indirectly prevented the accident. The company policy of optimal resource management would have provided the needed financial resources for adequate crew training. The company organizational climate that encouraged flight operation managers to do their best to make crew members maximally proficient would have had a positive effect on crew training. The encouragement of teamwork between crew members would also have changed their attitudes to all events occurring on board.

Pilot's Priority List: Teamwork

Since its first days, aviation has been a collective enterprise. It is physically impossible for only one person to successfully solve all problems connected with even normal flying of a commercial airplane. As in any complex system, small and big failures may occur in aviation. Although all possible failures are analyzed and alternate courses of action are developed during airplane design and operations, the number of required preventive and corrective actions dramatically increases in case of a failure. Only a well-trained team of professionals can be able to operate the airplane in all designed conditions.

Automated aircraft operations require even more teamwork. Cockpit crew and cabin crew have a lot of tasks that can be successfully completed by pilots or flight attendants if both groups of specialists act in unison, as a team. More than that, the entire aircraft crew must act as a united team. This condition is absolutely required in normal flights and especially in case of emergency.

Pilots and flight attendants will greatly benefit themselves and their colleagues as well as many other people if they remember to follow the requirement to act as a unified team from the moment they enter the company briefing room before departure until they say good-bye to each other after the flight.

9
Securing Crew-Aircraft Automated System Efficiency

9

Flight efficiency and flight safety are closely related. The more attention is paid to possible ways of increasing flight safety in automated aircraft flight operations, the more efficient these flights are going to become.

To operate automated aircraft safely, the crew members should be physically and mentally fit and professionally trained. The interface between pilots and the aircraft should correspond to human-operator natural abilities for perception of flight-relevant information, producing the required control inputs, and obtaining clear feedback from controlled systems. The flight environment should be organized in a way friendly to pilots' unerring activity. The aircraft automation should help pilots use the airspace in the safest way. From this perspective, some desired improvements in crew teaching and training, automated aircraft design, and airspace environment automated utilizing are suggested in this chapter.

The disasters that occurred with automated aircraft discussed in the preceding chapters were caused primarily by incorrect actions of the aircraft crews. Various conditions have led to these incorrect actions, but all of them were connected with crew-aircraft system qualities that have to be improved: insufficient pilot knowledge of aircraft design, pilot failure in monitoring important aircraft systems, inadequate pilot reaction to warning system indications, incorrect utilization of automatic systems, unsatisfactory crew coordination, and other similar shortcomings. It is obvious that to eliminate or at least to neutralize those shortcomings, the most urgent improvements are needed in three areas of the crew-aircraft-environment system characteristics that are connected with basic aviation skill and professionalism as well as with specific conditions of automated aircraft flight operations:

1. Something could really be wrong in the crew-aircraft systems that have been involved in accidents. It would be beyond human morality to blame our colleagues who unfortunately paid with their lives for their errors. But sometimes when a failure occurred aviators were not ready to act in a manner adequate to the automated human-machine system operation. Probably not all aviators who have encountered disasters knew about some negative factors that really exist in flight operations, and that these factors may reduce the pilot's ability to perform a safe flight even in normal conditions. An urgent need exists to tell pilots about those negative factors, as well as to provide them with proper professional instruction.

2. Flight safety records show that sometimes it is difficult for the flight crew to learn and understand how the automation actually functions, or how to operate it correctly, or why the pilot must not lose even a fraction of a second to perform an escape maneuver required by an automatic warning system. To facilitate the pilot's

tasks of learning automation during training and understanding it during flight, aircraft designers, together with experienced automated aircraft instructors, should confer and discuss how to expand their understanding of human limitations and how this can be factored into improvement of aircraft automation design and tailoring of flight crew training programs.

3. Flights are usually performed in airspace filled with other aircraft. The number of aircraft that may appear simultaneously in the same airspace area continuously increases. Aircraft automation has significantly changed the manner in which this airspace is utilized. Sometimes the influence of these changes may produce problems that were never seen before. This factor also must be accounted for in efforts to enhance automated aircraft flight safety.

Improvement of Pilot Literacy in Human Psychophysiology

Multiple statistical data confirm the fact that erroneous actions of aircraft pilots pose the main potential threat to flight safety. Known causes of pilot errors in aviation include

- Ergonomic factors
- Insufficient skills
- Encounter with unusual conditions
- Physiological factors
- Psychological factors

Ergonomic factors in flight operations are connected with insufficient aircraft suitability to provide a reliable human-operator activity. Ergonomic studies have played an important role in automated aircraft design since the very first days of automation implementation in aviation. Although it would be incorrect to fully deny the role of ergonomic factors in causing pilot error, the ergonomic research has produced significant positive results in modern aviation.

Insufficient pilot skills and inadequate training continue to contribute to the problem of pilot error. In the author's opinion, as long as aviation exists, this category of causes of pilot error will be at the center of attention among aviation academic circles, flight training institutions, and flight instructors. Later in this chapter we discuss existing and future approaches to pilot training in more detail.

The role of the cockpit crew as the leading link in the crew-aircraft system has dramatically increased as a result of the introduction of automated aircraft. The size of the cockpit crew has been reduced to just two pilots, while the volume of tasks that have to be controlled by human operators in flight has been left the same or even increased. For this reason, the reliability of pilot activity in flights has become even more important.

The transition of aviation from flying conventional airplanes with old-fashioned instruments and navigation equipment to flying modern automated aircraft has brought with it new flight crew tasks that require from pilots not only good basic skills and professionalism but also deep understanding of the processes of automation operation. Pilots' encounters with unforeseen operational conditions can result in their inability to immediately perceive signs of aircraft status changes, correctly process the new information, and make the right decisions that would restore the normal flight or would abort a potential emergency by correcting an existing failure.

The condition of a pilot as an operator in a human-machine system is characterized by a set of the pilot's physiological body parameters and psychological mind parameters. Optimal values of those parameters that guarantee reliable functioning of pilots as operators in flights are provided by various medical tests and other healthcare measures. Sometimes the parameter changes become so significant that they may reduce the pilot's ability to perform professional functions, or they may cause a pilot error.

Results of multiple research studies, including the author's work (Risukhin 1988), have proved that crew member efficiency significantly changes during every flight. The role of crew fatigue in flight safety attracted more attention in the late 1990s (Chittick 1998). Human body physiological and psychological parameters changes in flight may significantly impact crew efficiency and cause crew errors (Kozlov et al. 2000).

To overcome a specific problem, the first important step may be to learn about it. The pilot's physical ability to do the job in flight depends primarily on the pilot's individual body physiology, while the quality of the pilot's decisions is defined mainly by the pilot's psychological condition. Pilots' knowledge about physiological and psychological factors that are able to reduce the quality of their professional activity may better prepare them for possible difficulties and enable them to more correctly assess and compensate for their own and their colleagues' physical and mental conditions in flight.

Flight-relevant factors influencing human physiology

The pilot's ability to properly perform professional duties in flight, or pilot efficiency, depends on the intensity of specific factors inherent to flight activity. Many flight-relevant factors influencing the human body are known, such as air pressure, vibrations, and temperature. In normal operations of modern automated aircraft, the intensity levels for some of those factors can be considered acceptable from the human-operator activity point of view. But certain factors can really aggravate a pilot's activity in flight; these include duration of flight duty, sleep patterns, number of crossed time zones, monotony, and stress. The intensity of all these factors defines the human's physiological and even psychological condition in flight. The influence of these factors may result in a dangerous condition called crew member *fatigue*.

Fatigue

Fatigue is a combined set of changes in the psychophysiological state of the human body. It is a natural reaction of the human body to negative factors connected with flights. From the very beginning of automated aircraft operations with the reduced number of cockpit crew members, pilot fatigue became one of the major flight safety problems. According to the U.S. Aviation Safety Reporting System (ASRS) data, between 1980 and 1984, 261 incidents were recorded in which pilot fatigue was a contributory factor. The fatigue was also listed as a factor in 52 incidents recorded in the United Kingdom in 1983/84 (NASA 1986).

A definite level of fatigue follows any kind of human activity. As long as a person gets sufficient rest periods between periods of fatigue-producing activity, the fatigue stimulates the restoration processes in the person's body. But if the rest periods are too short or the negative factors are too intense and their influence lasts for a long time, the human body becomes unable to fully restore its functions. In this case the fatigue becomes cumulative. This level of fatigue must be avoided because it significantly reduces human professional abilities and can cause negative changes in human health.

Reduced pilot motivation for flight activity is one sign of cumulative fatigue. Intensive rehabilitation measures may be needed to return the pilot to normal psychophysiological condition. Timely recognition and reduction of factors that may cause negative changes in a pilot's physiology can control the fatigue level.

Flight duty duration

The professional efficiency of a pilot is not constant during the flight. Several phases of pilot efficiency in flight may be discerned. For instance, the efficiency of a pilot who had a good rest before flight progressively increases during the first 1 to 2 h of the flight. After that pilot efficiency remains at an optimal level for 2 to 3 h. The first signs of the pilot's tiredness emerge after 4 to 5 h of flight. At this period, which can last for several hours, the efficiency level does not decrease, because the pilot's body activates compensation functions. Then a phase of unstable activity begins. In this phase the pilots' efforts reduce, and errors begin to appear. A short phase called *final impulse* follows the unstable activity phase. And finally, a progressive reduction of efficiency begins; signs are a feeling of tiredness, fatigue, and apathy.

Errors caused by extreme flight duty duration can be avoided by optimal flight duty planning and maintenance. Automated aircraft crews often perform flights that continue for 10 h and even longer. In this case a pilot efficiency retention strategy must be used. This strategy is based on including in the cockpit crew one or even two additional crew members called *relief pilots*. Normally a relief pilot has the same rating as the aircraft captain. The function of the relief pilot is to temporarily replace the main crew pilots in turn to allow both of them to briefly rest during the flight. This measure significantly improves crew efficiency by refreshing the pilots and helping them perform correctly during the last period of flight.

Sleep quality
The human body restores itself during sleep. A sufficient amount of sleep before departure is critical for maintaining an acceptable pilot efficiency level in flight. To avoid reduction of working efficiency and errors in flight caused by unsatisfactory sleep, pilots have to properly organize their activity and nutrition before the sleep time. Even several hours of sleep before flight at any time of day can produce a positive effect on the pilot's performance.

Number of time zones crossed
Long flights of automated aircraft inevitably cause their crew members to visit Earth time zones with local times that may differ from the crew's home local time by 12 h. Because all biological processes are based on time (circadian) rhythms, the frequent time zone changes may have immediate negative consequences on crew members' activity in flight and more distant negative consequences on their health in future. This effect is often referred to as "jet lag."

The intensity of jet lag also depends on flight direction. Flights from east to west generally cause less discomfort for an average person in comparison to eastbound flights.

Another factor that influences jet lag is the degree of coincidence of the pilot's body biological rhythms with the flight schedule. For example, departure from Moscow, Russia, to New York at 10 A.M. Moscow time is quite easy for Boeing 767 Russian pilots' bodies because it fully coincides with their biological rhythms. After 10 h of flight they arrive in New York, where it is noon. The crew members perform approach and landing at JFK (John F. Kennedy International Airport) in a relatively good working form because their bodies' biological clocks show only the last hours of wakefulness.

But for their Russian colleague crew members who leave New York at 3 P.M. U.S. Eastern Standard Time, the departure will be at 11 P.M. of their home (and body) time. To be in good working form before departure, they are strongly urged to wake up early in the morning in New York and have 2 to 3 h of sleep immediately before leaving their hotel. During flight over the Atlantic Ocean the pilots' bodies will naturally want to reduce their activity levels. At this period pilots have to mobilize all their will and skill to stay in the aircraft control loop and to correctly perform all duties and maneuvers needed for the transatlantic flight. A short rest, made possible by the third pilot (a relief captain), taken by both of the pilots who are going to land at Moscow at 8 A.M. local time will help them reliably cope with this task.

Monotony

The factor of monotony in flight is caused by uniformity of pilots' actions, their isolation from the rest of the world, and even the sounds in the cockpit. The monotony can result in reduction of pilots' psychological activity, which in turn may cause extreme relaxation, forgetfulness, and negligent actions.

To avoid monotony, it is recommended that pilots periodically perform physical exercises during flight. Increased levels of cockpit lighting can also be helpful in overcoming monotony.

Stress

The stress phenomenon is a human body reaction caused by strong irritation. The irritation can be expressed in physiological as well as psychological forms. In aviation psychological stress is more common.

The stress reaction may significantly worsen the operator's qualities required to control the aircraft flight. The pilot's visual field constriction, reduced memory, and changed movements can all be caused by stress. Decision-making abilities are also negatively influenced by stress. Consequences of continuous stress may have a negative effect on pilot health.

To avoid pilot stress during flight training, the training programs have to be well balanced. Difficult elements of flight should be discussed, shown, and practiced by flight instructors in a calm and successive manner. Unexpected imitation of emergency situations in flight simulators can help to significantly reduce pilot stress during real flight operations.

Psychological factors

In the early 1970s the author used to fly to a small airport near a coal miners' town called Sangar in central Yakutia, Russia. There was a mountain about 3000 ft high located along the airport runway axis several miles away from the airport. A road ran to the top of the mountain. A nondirectional (radio) beacon (NDB) located near the airport control tower was used for instrument approaches. In good weather conditions visual approaches were made along a pattern rather close to the mountain, and aluminum fragments of an airplane were seen through pines on the mountain's slope. Older pilots stated that the fragments belonged to an airplane similar to the DC 3 that collided with the mountain while making a nighttime flight. During the NDB approach the airplane pilots mistook several lights along the road on the mountain for the runway lights. Although all instruments, including the automatic direction finder (ADF) and artificial horizons in the cockpit, operated normally, the pilots did not believe them and

Securing Crew-Aircraft Automated System Efficiency 271

as soon as they saw the road lights, they turned the airplane directly into the mountain.

This story confirms that in pretty early aviation times pilots did not fully believe the airplane instruments and that this mistrust cost some of them their lives. Nowadays an opposite event sometimes occurs—pilots rely on aircraft automation too much, do not properly monitor the aircraft systems themselves, and also lose their lives because of their blind faith in automated systems infallibility. The A310 aircraft accident that occurred during takeoff in Bucharest, Romania (case 2, described in Chap. 2) is one of several tragic examples. During takeoff the pilots believed that the autothrottle system would control the thrust of both engines in the way that it was expected to do. But instead, the autothrottle failed to retard the throttle of one engine. Pilots were unable to recognize the failure in time and lost control of the aircraft. Both deadly errors, one that occurred several decades ago and another that occurred relatively recently, can be explained, in addition to other factors, by one natural thing: pilot psychology.

Modern aviation indeed recognizes psychological phenomena that can be dangerous for flying. Negative psychological factors that can cause pilot errors really exist and have to be recognized by pilots. This knowledge is especially important for pilots of automated aircraft, because in these aircraft mental activity represents the biggest part of the pilot's job, and because the price of a cockpit crew member's error in these aircraft is extremely high. Although all the negative psychological factors that may interfere with the flight safety cannot be described in detail in this book, in the interest of the reader's safe flying at least some of those factors seem worth mentioning and briefly defining.

Illusion of located target

This psychological phenomenon can be illustrated by a landing on a wrong runway. When approaching the airport in adverse meteorological conditions with low visibility, a pilot needs to find the airport runway as soon as possible. Insufficient visual information perceived by the pilot is supplemented by the pilot's imagination. Objects seen by the pilot that are unrelated to or not part of the airport are interpreted as signs of approaching the desired runway. Also, the pilot may land on the wrong runway, believing it to be the correct one because it is located in a direction similar to that of the target runway.

To prevent errors caused by this factor, called *illusion of located target*, the pilot must identify all symptoms and signs taken into consideration for navigation situation awareness. Perception and understanding of radio navigation data indicated in the cockpit can be extremely helpful in this situation.

Dominant condition

Sometimes pilots forget to retract speedbrakes during final descent before landing, or they do not react to the GPWS signals by immediate transition to the aircraft climb. Although all pilots are practically never motivated to commit an accident, and these errors may lead to catastrophic consequences, pilots in most cases cannot explain why they acted in that incorrect and even dangerous way. The true roots of these kinds of errors are also contained in the pilot's psychological makeup.

At any given moment the psychological condition of a human being consists of parameters such as attention, thinking, will, and emotions. To provide optimum performance in flight, all these qualities of a pilot must be balanced. If for some reason the regulation center of a single flight task becomes much stronger than others, the pilot stops to perceive information concerning other important tasks. This psychological factor, called a *pilot dominant condition*, can be caused by various social, professional, or medical reasons. The pilot's desire not to violate the company crew rest regulations by landing as soon as possible in an unclear navigation situation is an example of this factor.

To avoid errors caused by dominant conditions, pilots are advised to achieve professionalism in distributing crew member attention in flight; distributing operations, tasks, and functions within the crew in an optimal way; continually providing reciprocal checks; and avoiding situations that may exhaust the pilot's abilities.

Drowsiness

In this condition a crew member may perform actions that are fully irrelevant to the flight phase. In some known cases, after takeoff drowsy crew members have retracted aircraft flaps instead of landing gear, stopped all engines in flight, and performed other dangerous actions. Drowsiness is a transition condition between staying awake and sleep. The danger of this condition for aviators consists in the fact that the person's motion activity is retained while this activity is controlled by dream images. Reasons for this condition include fatigue, shift of biological (circadian) rhythms, and monotonous environment.

Crew errors caused by drowsiness can be avoided if aviators' work, rest, and nutrition processes as well as their sleep and wakefulness patterns are optimally organized, and if crew members communicate and check each other's activities during flight.

Deliberate distortion of motivation

A hierarchy of crew targets exists in every flight. This hierarchy includes

1. Flight safety

2. Comfort of passengers
3. Operation schedule
4. Company economy

Normally motivated crew members act in a way that provides achievement of the targets in accordance with their hierarchy. But sometimes positions of targets in this hierarchy may be deliberately changed by flight operations managers or by crew members themselves. For example, financial reasons or personal comfort may suddenly be considered more important than flight safety priorities. In some cases this change of priorities may cause a disaster. An aircraft accident in the airport of Ivanovo (Russia) occurred because the captain did not want to miss a bus that was to carry the flight crew home after arrival. An approach to land at a well-known airport was performed. The crew members were fully aware of the flight situation. The captain, who was trying to land as soon as possible, disregarded other crew members' cautions about violating the approach procedure. Making a visual approach instead of a recommended instrument approach, the captain tried to perform aircraft evolutions at a low height. As a result, the aircraft collided with a building in the airport vicinity.

This psychological factor is called *deliberate distortion of motivation*. To avoid errors caused by this factor, crew members must be properly motivated during the entire flight. The company policy should encourage priority of flight safety and flight discipline among all employees connected with flight operations.

Failure to switch for a new activity

There is another psychological phenomenon: the closer a person approaches to a target, the stronger the motivation to reach that target. This phenomenon is known as *failure to switch for a new activity*. In aviation, examples of errors caused by this factor are failure to perform a go-around maneuver after suddenly encountering poor visibility and failure to proceed to an alternate airport while the fuel amount on board still allowed that maneuver. The reluctance of automated aircraft crew to take over a malfunctioning system can also be considered a failure to switch for a new pattern of actions in time.

Errors of this sort can be avoided by careful analysis of possible situations, and by preliminary planning of alternative crew actions in case the flight situation is aggravated.

Psychological direction

The pilot may be directed to a definite sort of information or activity. If this direction is sufficiently strong, it can override incoming information about new emerging conditions, which require changes of planned actions. For example, a pilot, who had rating only for visual flights, decided to fly a small airplane with his closest relatives on board to a family celebration. Although en route he encountered complex meteorological conditions that did not allow visual flying, he did not change his initial plan and continued the flight until he lost the space orientation. The airplane crash was the result of the pilot's strong psychological direction to complete the planned flight. This desire did not allow the pilot to assess the new meteorological situation and to change the flight plan accordingly.

The author was a member of a team that investigated another accident caused by the pilot's psychological direction. The captain of a four-engined turboprop cargo aircraft decided to perform takeoff from a runway covered with a wet snow layer of 3 to 5 in, which exceeded a maximum allowed amount of runway contamination. The aircraft could not achieve the acceleration needed for liftoff and collided with obstacles outside the airport. All people on board died because of the captain's desire to perform takeoff notwithstanding the unsatisfactory runway condition.

Clear understanding of the flight situation and effectively arranging the flight safety priorities can help avoid pilot errors caused by the psychological direction factor.

Complacency
Pilot complacency is another negative psychological factor that became common in automated aircraft flight operations. High reliability of modern aviation automatic systems causes some pilots to believe that no failure can occur in their flight. After many dozens of flights completed without any failures, the pilots become psychologically immobilized to timely and correctly discover and localize a system failure or a navigation error.

Systematic full flight simulator training that imitates aircraft systems failures in real flights can be useful in avoiding crew errors caused by pilot complacency.

Securing Crew Proficiency

Even excellent conditions of the pilot's body and mind do not guarantee timely and correct actions in normal as well as in abnormal flight situations. Training pilots to proficiency is an absolutely necessary condition for efficient flying of automated aircraft.

Crew training varieties

Pilot proficiency in aircraft automation utilizing is based on two attributes: ability to operate the aircraft in flight as a professional pilot, and ability to perceive, understand, and correctly react to specific automation operational signals.

These days, automated aircraft pilots are seldom trained to fly these complex machines immediately from the very beginning of their pilot careers. Normally they obtain initial flying proficiency while flying airplanes equipped with conventional instruments and flight-control systems. These airplanes are much less automated, so it may be suggested that pilots now gain genuine flight handling skills well before they begin to operate automated aircraft. This suggestion allows them to pay most of their attention to the second set of pilot qualities needed for proficient automated aircraft flying.

Three kinds of automated aircraft pilot training are used at the present time: transition flight training, recurrent training, and crew resource management (CRM) training.

The pilots' abilities to perceive information relevant to automated system operation, to understand it, and to make correct decisions and actions depend primarily on the pilots' knowledge of the aircraft design and operation and the pilots' ability for teamwork in flight. This knowledge and ability are supplied and maintained as a result of the pilots' training.

Transition flight training
Airline pilots' transition training for automated aircraft is provided in a well-structured form. It consists of a ground course, fixed-base simulator training, full flight simulator training, and the airplane training.

The ground course is aimed to provide the pilot with the minimum aircraft knowledge needed for reliable flight training and further independent flight duties. Normally this part of the training process is computerized. Each pilot is supplied with a personal computer, its software, and hardcopy literature utilized in accordance with the training program. A step-by-step learning method is used. Learning a specified amount of material is followed by control questions that, if answered correctly, allow the trainee to pass to the next lesson.

Fixed-base simulator training is designed to develop pilot skills in preparing aircraft systems utilization and in systems operation in flight. A *fixed base simulator* is a motionless computerized copy of the learned aircraft cockpit. During this part of training pilots are also introduced to crew coordination principles used in automated cockpits.

Full flight simulators are used to develop the basic aircraft flying skills in pilots. Acting pilot-instructors of the particular aircraft type perform the training. Usually the same instructors provide pilots with the real airplane training.

This training structure can supply new pilots for automated aircraft with the required professional qualities. To have employed well-trained pilots able to fly the airline's aircraft safely, the airline has to satisfy three conditions: the needed level of trainees' professional background before the training, quality of the training programs, and flight instructors' professionalism.

Recurrent training

Automated aircraft in general as well as their systems are reliable and redundant. Normally the first failure of a system has a rather small probability of occurrence, and it does not result in an aircraft emergency. The probability of the second failure of the same system is even smaller, but the consequences may be more serious, and localization of the failure requires crew actions different from those used in normal flight operations.

To keep pilots always ready to act correctly in case of emergency, the airlines use *recurrent* training, which is provided on average every 6 months. This training is very similar to the transition full flight simulator training, with emphasis on correct crew actions in cases of engine failures as well as hydraulic, pneumatic, electric, and automatic control failures. Scenarios of engine and aircraft fire, rapid cabin decompression, emergency descent, and passenger evacuation are also mandatory elements of the recurrent training. And again, the quality of training depends on programs and instructors. These two training attributes are rather expensive, but they are the only real tools for achievement and maintaining airline pilots' proficiency and airline flight safety.

Crew resource management training

Crew resource management (CRM) training became a recognized tool for controlling crew error. Nowadays, CRM training can be defined as a human-factors-based discipline, "which is concerned with human-machine and human-human interfaces" (Helmreich 1999). Training crew members to work in flight as a united team is the core CRM destination. Colleague crew members, air traffic controllers, maintenance teams and other ground specialists, and the aircraft, with its systems, operations manuals, navigation documentation, and many other objects, are resources that the crew can use to successfully complete the flight. CRM training is aimed to teach crew members to use all the resources as well as to develop abilities in leadership, situation analysis, and decision making. Training pilots to make timely and correct decisions is one of the most important CRM training objectives.

Although CRM training has not totally removed human errors from aviation, its results have allowed significant improvements in flight safety. For example, the United Airlines' accident rate decreased by about 5 times since the airline began its CRM training program (Krause 1996). The leading airlines use CRM training for cabin crews, maintenance teams, and other aviation professionals, and this acronym is now interpreted as *company resource management*.

For a long time, since they emerged in the early 1980s, CRM training programs did not depend on aircraft models operated by trained crews. Usually an airline used one training program for crews of all aircraft that were utilized in the airline. Because crew coordination and communication in automated aircraft differ from their conventional counterparts, it became evident that aircraft type-specific CRM training programs would be more appropriate for crews that operate glass cockpit aircraft (Wiener 1993). It is expected that these programs will produce a positive effect.

Additional directions of crew training

To provide reliable flight operations of new aircraft in the twenty-first century, additional directions in flight crew training activity seem to be required. These directions may be described as follows.

Automated aircraft crew training philosophy

A crew training philosophy tailored for automated aircraft flight operations is needed. Pilots' understanding of automation and their ability to manage the automation and to decide what levels of automation are appropriate in specific circumstances proved to be critical.

The training should be sufficient to prevent pilots from inconsistencies and uncertainties in utilizing automation. As a result of training, pilots should use automation effectively and should be prepared

- To understand automatic device design, operation, and interaction
- To monitor automation effectively
- To understand functions and limitations of automation modes
- To rapidly intervene in an automated flight-control process

- To fully assume the aircraft control from automation
- To recognize approach to stall and other marginal conditions
- To perform without delay escape maneuvers in any case of warning activation
- To plan and control the automated flight using the aircraft's flight management system
- To make decisions with the help of automation, and not to allow the aircraft automation to make decisions

Modern aviation requires that pilots, in addition to having the ability to operate computerized aircraft control systems, maintain all qualities that constitute "airmanship": discipline, skills, knowledge, situational awareness, and making decisions (Kern 1997). New training programs should provide pilots of automated aircraft with these qualities.

The new crew training philosophy should take into account that safe operation of modern aircraft in flight depends not only on pilots but also on flight attendants as well as other people (e.g., ATC) outside the aircraft during flight.

Automation training requirements for instructors and check pilots should be developed. The requirements should emphasize the senior pilots' abilities to explain and assess computerized aircraft functions and to show the trainees how to utilize those functions.

Selection of candidate pilots to be trained for automated aircraft should provide for choosing candidates with adequate professional background for the new job, and with physiological and psychological suitability. There are signs that learning to use automation may be more difficult for older pilots as well as for captains of conventional aircraft that are controlled by crews, including a navigator and a flight engineer. Preliminary English language training before the training course for pilots with native languages other than English is strongly recommended. Classes on automatic control principles can help trainee candidates to subsequently understand automatic aircraft device design and operation.

Closed-loop pilot training

In spite of the significant amount of pilot training used by airlines, pilot errors similar to those that occurred previously continue to occur. The reason for these unsatisfactory results can be explained by the fact that the training often is an open-loop process. Quality of training becomes obvious only after periodical pilot checks in flight or in simulators by check aviators. Incident and accident investigations also provide information about pilot performance, but they cannot be considered a desirable feedback for pilot training assessment. The task of introducing closed-loop pilot training with the possibility of immediate training results assessment has been discussed by training specialists (Small et al. 1999). The idea is based on using flight and simulator data analyses for assessment of the training effectiveness.

Decision making training

Decision errors cause most flight accidents. Understanding of the situation and an appropriate response are important elements for reducing decision errors. Gaining flight experience and exposure to various flight situations usually improves the pilot's decision-making abilities. But sometimes the price of this learning is too high. Nevertheless, training pilots in making correct decisions represents a comparatively small share of pilot training programs (O'Hare and Roscoe 1990).

The increased role of decision making in automated aircraft crew activity and the significant amount of decision-related crew errors suggest that decision making training should become an independent training discipline. The following steps are recommended in decision-making training (Orasanu 1993):

- Understanding the problem before acting
- Assessing risk and time factors
- Developing response strategy based on situation features
- Establishing contingency plans
- Considering less obvious future consequences
- Managing workload
- Communicating the problem with other involved people (cockpit, cabin, and ground personnel)
- Defining the plan, actions, and assignment of responsibilities

The pilot's participation in completion of these steps, even for making decisions in less demanding situations, may be helpful for developing a stable pattern of the decision-making process.

Crew-Aircraft System Optimization

Automated flight operations efficiency also depends on an optimal completion of processes relevant to the whole crew-aircraft-environment system. These processes include aircraft crew flight activity management in the airline, reciprocal accommodation between the aircraft and its crew, and enhancement of the utilization of the airspace environment by the crew-aircraft system.

Crew management

Crew management policy established in an airline influences that airline's crew efficiency through approaches to flight crew staffing practice, ensuring that pilots remain professionally current, and standards of aircraft automation utilization in flight.

Crew pairing

The automated aircraft cockpit crew consists of two pilots. At least one of them should have automation operations experience that significantly exceeds the minimum required by company regulations. This precaution can compensate for some invisible deficiencies in professional training of an individual pilot. The deficiencies may be difficult to discover during check flights after transition training, and they may show up in the new pilot's first independent flights. Even the very presence in the cockpit of a more experienced crew member can be very helpful for the younger pilot's further professional growth.

Pilot compatibility is another factor to be taken into account by flight operations managers. As there are no two absolutely equal people, there can be no two professionally equal pilots. The flight profession requires a lot of positive qualities from a pilot candidate. It is difficult for an average pilot in real life to constantly maintain all those qualities at a high level. Sometimes even very good pilots may have temporary personal or professional difficulties. The art of a crew manager is contained in recognizing those difficulties in a timely manner and compensating for them by assigning a wittingly reliable pilot in the cockpit.

Mixed-fleet flight operations

These days some airlines use pilots flying simultaneously on several types of aircraft. There are two reasons for this approach: (1) This policy helps airlines in fleet crew staffing, and (2) aircraft building companies have significantly standardized and unified cockpit layout of different aircraft types produced by the same company. The families of Airbus and Boeing aircraft can be good examples of that similarity.

Possible problems connected with this operational approach have to be mentioned. Simultaneous flying of conventional and automated aircraft may negatively affect a pilot's skills to operate aircraft of both types. Conventional aircraft flying requires more manual control inputs and mental flight path calculations, while automated aircraft are controlled mainly via complex autopilot flight director system modes and assessment of automation operation, which consumes a significant part of the pilot's attention. When the pilot operates one type of aircraft and then another, this may result in confusion in flight-control priorities.

Flying of automated aircraft made by different corporations may also degrade a pilot's abilities to adequately control the aircraft. This may be caused by different cockpit layouts as well as by differences in automatic flight mode names and indications. Different methods of pilot control inputs via control columns or sidesticks during manual control should also be considered.

One aircraft manufacturer's automated aircraft flying seems to be the most appropriate way of simultaneously using pilots to fly two aircraft types. But even in this case crew errors are possible, because of visible or hidden differences between aircraft cockpit hardware and software.

To avoid pilot errors introduced by the effect of simultaneously operated aircraft, recurrent ground, simulator, and airplane pilot training for both aircraft types should be conducted systematically in accordance with regulations established by the state aviation authority.

Pilot professional currency

The relative complexity of automated aircraft operations in flight in comparison with conventional airplanes requires a more careful approach to ensuring that pilots remain professionally current. Long time intervals between flying automated aircraft may cause pilots to lose their automation utilization skills. If a pilot had a significant (more than one month) break between flights, the first flights after the break should be performed under a senior pilot's supervision, and the pilot's review ground training should precede those flights. Additional recurrent flight simulator training may also be needed to restore a pilot's automation skills.

If pilots fly continually, the airline policy should encourage them to control automated aircraft in both automated and manual modes.

Pilots operating simultaneously two or more types of aircraft must have recurrent training on all the types in accordance with airline policy.

Automation utilization standards

The company standards established for flight crews in aircraft automation utilization may also affect flight operations efficiency. Although companies producing automated aircraft develop variants of automation that they consider optimal for use by the flight crew, airlines that operate the aircraft usually develop their own automation utilization standards. This development is intended to tailor the automated aircraft flight operations in accordance with the national aviation legislation, the airline policy, its flight personnel's professional background, and other factors.

Sometimes company standards require the use of aircraft automation during all phases of flight. This requirement may not be clearly defined, but the vast number of tasks that pilots should perform in flight forces the flight crew to use automation whenever possible. Tasks that are not directly connected with flight safety include company commercial radio communications, filling in supplementary information documents and forms, and unusual requirements for aircraft captains' comments for passengers about the flight route. In these cases pilots slowly lose their proficiency in manual aircraft flying. In some circumstances this may also lead to pilot error.

Flight procedure design is another source of possible crew errors. Sometimes aircraft maneuver design in a specific airport area may require, for correct completion, mandatory use of automation. In other cases abrupt changes in aircraft maneuvers or changes in the aircraft flight path at a low altitude are included in the departure or approach maneuvers. These unnecessary complications should be avoided because they significantly increase crew workload and are potential sources of pilot error.

Crew-aircraft adaptation

The efficiency of every flight is determined by its economical results and its safety record. Positive economical results of the use of automation in air transportation have been proved by the history of automated equipment expanded design, production, and operations in this industry. Use of aircraft automation has allowed significant reduction in flight operation costs. However, to avoid compromise in flight safety, an optimal balance between commercial and safety issues in aircraft automation design should be maintained.

In automated aircraft, the pilot's abilities to perceive and understand information relevant to the aircraft flight path parameters as well as the aircraft systems operation data, to make correct decisions, and to complete needed actions considerably depend on factors defined by the aircraft design. The experience acquired in the aviation world in automated aircraft flight operations allows some suggestions directed at better adaptation between the flight crew and the glass cockpit (Funk et al. 1996). For a long time in the aviation world, despite improvements in overall aircraft design, airplane crews usually had to continue their operations in older-style cockpits, because of the high cost of airplane cockpit design changes and insufficient knowledge of human-machine system performance in aviation; these days, however, when new automated aircraft types and models are created, the aircraft cockpits may be designed or modified to increase efficiency of their crews.

Crew-automation interface improvement
The interface between aircraft automation and flight crews proved to be extremely important for successful flight operations of automated aircraft.

To provide efficient crew activity in flight, all flight-relevant information should be presented in the cockpit in accordance with inherent human perceptual and cognitive abilities. Size and quality of images and symbols should not impose additional difficulties in information perception.

To avoid errors in utilization of automation modes by pilots after their transition training from other automated aircraft, industry standards are needed for mode marking and indication, as well as for standardization of aircraft automation systems function. Standards also should be developed for the displays and controls on flight decks (ITSA 2000).

To avoid unnecessary increase of pilot workload, which could result in pilot errors, the cockpit design should factor in pilots' intuitive skills and previous flying experience. Aircraft control strategies utilized by

Securing Crew-Aircraft Automated System Efficiency 289

automation should be similar to those of a human pilot.

Extreme simplification of the pilot-automation interface may produce the blackbox effect, where the pilot does not perceive and understand even the principal functions of a controlled system operation. To provide the pilot with cognitive references during operation of an automatic control device, the complexity of the device interface should be balanced with its functions.

New cockpit indication introduction

Flight safety data analyses have shown that some additional indicators of flight path parameters and aircraft controls are desirable in the cockpit.

The *angle of attack* (AOA) is the most important flight parameter that defines the very ability of the aircraft to provide its basic function—to fly in the air. Many conventional airplane pilots saw this parameter displayed on the instrument panel. Experience has proved that the absence of this indication in the glass cockpit cannot be justified by the automatic flight envelope protection functions provided in modern aircraft. If pilots involved in some recent aircraft incidents and accidents had had access to AOA displays and were properly trained to use this indication, the consequences of those events could have been much more favorable.

There is one more flight path parameter indication problem that, in the author's opinion, is extremely important for flight safety in general as well as for automated flights in particular. It is the way in which *aircraft bank attitude* is presented to pilots. Two methods of this indication are known:

1. The first method is called "view from the aircraft to the Earth." This method (described in Chap. 4) is used, in particular, in automated aircraft as well as in plenty of

other airplanes. Its main feature is that the artificial horizon position relative to the aircraft at any moment of flight fully coincides with the natural horizon position seen by pilots from the cockpit. The aircraft bank attitude side and approximate value are indicated on the attitude indicator by the relative positions of the artificial horizon and a motionless aircraft symbol. Exact values of the aircraft bank are indicated on an outer scale of the altitude indicator by the end of a vertical line (the pitch scale), perpendicular to the artificial horizon.

2. The second method of airplane bank attitude presentation is based on the use of a movable airplane symbol that indicates airplane bank attitudes relative to a horizon line imitation, which at any moment is parallel to the airplane's lateral axis. In this case the artificial horizon in the cockpit does not coincide with the natural horizon. This fact is used mainly to justify the first method of bank indication and to reject the second one.

However, results of aviation scientific research (Ponomarenko and Zavalova 1992) have proved that the pilot of a modern all-weather aircraft does not immediately use the natural horizon position or its indication for controlling the aircraft bank. An image of the flight exists in the pilot's brain, and the cockpit bank indications, together with other indications, continuously update that image. The bank indication perceived by the pilot by method 2 is fully appropriate for immediate use by the pilot's brain because the pilot obtains the aircraft bank side indicated by the aircraft symbol in the same way that the pilot perceives the flight image. Bank indication method 1 causes the pilot every time to recode visible indications of the bank into a form acceptable by the mental flight model. This action always requires some time and increases the pilot workload. To avoid pilot errors caused by incorrect interpretation of aircraft bank indications, attitude indication method 2 may be recommended.

The next suggestion, proposed by the author, who is a Boeing 777 instructor pilot, may cause somebody to smile. But flight safety is a problem that requires consideration of any potentially useful proposition. An old-fashioned instrument that would not be unnecessary in the glass cockpit is a gyroscopic indicator of turn. Pilots of older airplanes were trained and could use it instead of the artificial horizon in instrument flight conditions. Of course, the reliability and redundancy of automated aircraft attitude indication is very high. But who can provide an absolute guarantee that no one automated aircraft will not be, at some unhappy time, totally deenergized in flight, for example, as the result of thunderstorm lightning? In this case the simple and not very expensive device fed by a single electric battery may save many human lives.

A well-noticeable indication of speedbrakes-deployed position can also be recommended for introduction in all glass cockpits. Some automated aircraft have this indication on a system display, but in critical moments of flight pilots may omit it.

Reversion to a lower-level mode of automatic control may disable the flight envelope protection function that preserves the aircraft from violation of aerodynamic limitations. If pilots utilize this mode rarely, they may erroneously rely on the disabled automation protection. For example, the airspeed protection function is not available in the vertical speed automatic flight mode. In cases where this mode is engaged, pilots should be reminded of it by a corresponding indication.

Automatic control system function modification

Automatic control systems of modern aircraft are safety-critical systems. Flight operations experience shows that some improvements in these systems may be useful in terms of flight safety.

Sometimes an automatic flight-control system maintains the flight path parameters well after the normal aircraft balance condition has been broken because of a serious failure. The autopilot that increases the wing angle of attack for maintaining an altitude that was maintained before the occurrence of an uncontrolled bank causing a descent, or the autothrottle that maintains a previously set speed after an engine failure by increasing thrust of a good engine are examples of how automation can aggravate the emergency situation. A device that could inform pilots about the dangerous flight-control situation may be an indicator that starts to show that normal automatic control inputs have been surpassed as soon as established levels of those inputs are reached.

Three features in automatic flight-control design should be discussed and, possibly, modified.

1. *Multifunctional use of the same automation control in the glass cockpit.* In some automated aircraft, to change the autopilot flight director mode from lateral navigation to heading select, the pilot must perform three actions with the same control: to push, pull, and then turn the heading-control knob on the automatic flight-control unit. A maximum of two actions with the same control may be more acceptable to avoid unnecessary complications in the pilot-automation interface.

2. *Reducing forces that may emerge on flight controls on autopilot disconnection and transition to manual control.* During flight, this feature may produce additional difficulties for the pilot, especially at low altitudes.

3. *The envelope protection function.* This function is widely used in automatic control systems, but its autonomy with respect to the pilot seems to be excessive. The

function should provide pilots with the ability to manually perform maneuvers outside the protected envelope that may be needed for recovery from unusual attitudes to avoid an accident.

Airspace utilization enhancement

Automated aircraft navigation systems are designed to utilize the airspace as a part of the flight environment in flights via air routes and in airport areas. Airspace utilization is organized in an economically optimal and simultaneously safe way. At the same time, practical flight operations have demonstrated possible ways for further improvement of flight safety in the process of airspace utilization by insignificant modifications in aircraft automation software and hardware.

"Blind automation precision" effect elimination

The lateral and vertical precision of automated aircraft air navigation is very high. On routes located in certain airspace areas, this precision, in particular, has allowed reduction of vertical intervals between aircraft to only 1000 ft. If the cockpit crew follows all required procedures and recommendations, the aircraft normally maintains an assigned altitude and proceeds practically along the air route axis.

But this excellent technological achievement may also potentially threaten flight safety in case two aircraft flying in opposite directions for some reason maintain the same altitude. The high speeds of aircraft flying at the same air route altitudes practically dismiss the possibility for pilots to rapidly divert the aircraft to avoid collision. Thus the negative effect of "blind automation precision" should be eliminated.

To avoid this potentially dangerous situation, a simple modification of existing flight management systems

(FMSs) can be suggested. The FMS software should be configured in a way that all aircraft in automated flights always are maintained at the assigned altitudes, but at some distance from their air route axes (e.g., 3000 ft) to the right within the air route lateral width. This additional precaution cannot worsen economic and navigation parameters of flights, but will provide an unquestionable additional flight safety guarantee.

Instrument approach equipment quick readjustment

In some airports, because of the dense air traffic, the flight crew, during approach to land, may obtain the air traffic controller's direction to change the runway intended for landing. This change may be issued on final descent of flight. Even if flight conditions allow visual flight, pilots should have instrumental flight path indications to provide the aircraft with optimal attitude at this important flight phase. Retuning of the aircraft instrument landing equipment takes at least several seconds and distracts the attention of one pilot from monitoring the flight path during this critical time.

To avoid pilot errors caused by these circumstances, automated aircraft equipment should enable the pilots to readjust the aircraft instrument landing system in a short (1- to 2-s) time period, without significant pilot distraction from the flight path control.

Terms of Safe Automated Aircraft Flights

Optimal actions of flight crews produce desirable results in flight safety. Effective crews properly trained to manage their resources deal proactively with flight safety problems. Their actions include reciprocal briefing on known safety threats; they ask questions, speak up, and constantly reevaluate their decisions; they clearly communicate operational plans, prepare and plan for safety threats, distribute the workload and tasks, and exercise vigilance; their captains show leadership (Helmreich 1999). If a severe threat to flight safety emerges, a properly trained and disciplined crew can cope with it and save the lives of passengers, crew, and the aircraft.

Case 9: Engine reverse-unlocking incident

On August 16, 1998, an A310 crew performed a scheduled flight from Tokyo, Japan, to Moscow, Russia. Before departure the mechanical condition of the aircraft, conditions at the departure airport, the weather, and other flight-relevant factors were normal.

After takeoff from Narita airport in Tokyo, the crew reached flight level 350 and performed a cockpit scanning procedure required by the airline regulations. No abnormalities were found, and the flight proceeded in accordance with its flight plan.

When the aircraft was over the Japan Sea 50 min after takeoff, a warning lamp illuminated signaling reverse unlocking of the right engine. The pilot not flying at once noticed the indication and informed the pilot flying about that. Both pilots checked the engine parameters, which were normal at that moment. The aircraft quick reference handbook (QRH), describing emergency procedures, was brought out for ready access in case it would be needed.

A special engine reverse-unlocking procedure was given in the aircraft flight crew operations manual. The crew should have applied this procedure in case a buffet or an aircraft bank followed the indication. No bank or buffet was observed in the initial moments after the right-engine reverser failure indication appeared.

The captain, aware of the extreme danger of the reverse-unlocking activation in flight, decided to reduce the right-engine thrust. As soon as the right-engine throttle lever was in idle position, the engine vibrated severely, and a right bank occurred. The pilots immediately performed the engine reverse-unlocking procedure.

The QRH as well as the aircraft flight crew operations manual required the pilot to move the failed engine throttle lever to idle position, move the engine's fuel lever to OFF position, maintain airspeed at or below 240 knots, and follow another, additional procedure for aircraft single-engine operation.

The first requirement of the single-engine operation procedure was to land as soon as possible. Because the nearest airport was Narita, the originating airport, the captain decided to return there. The APU (auxiliary power unit) was engaged, and other needed adjustments were made in a timely manner. The return (to Narita) one-engine flight was performed in accordance with corresponding rules and requirements.

Because a long flight was planned earlier, the weight of the aircraft before landing at Narita exceeded the maximum landing weight allowed for normal operations by 3.3 tons. Because the aircraft did not have a fuel jettison system, the weight could be reduced only by additional flying time in the airport area. However, considering the threat to flight safety imposed by the right-engine failure, the captain decided to perform a so-called overweight landing, which requires more precise flight-control input from the pilot flying during the very last moments of flight.

Disaster has been prevented

The aircraft landed safely at Narita 1 h and 57 min after departure.

Investigators later determined that the incident was caused by an improper maintenance troubleshooting of a failure in the right-engine reverse-control system.

Pilot's Priority List: Expanded Professionalism

The safe landing of the A310 flight at Narita was the logical result of skilled and well-coordinated crew actions during the flight. The crew's quick, effective actions even corrected the errors made by other people on the ground (mechanical technicians) who had failed to provide proper aircraft maintenance services. The pilots had immediately noticed the right-engine failure indication. Although no other signs of abnormal system operation were initially perceived and no crew actions were formally required, the pilots at once prepared themselves for actions in case the situation worsened.

The captain's precautionary reduction of right-engine thrust showed that the failure had really occurred, and the pilots immediately took corrective actions. Usually the very first moments of the engine reverse system failure are the most dangerous, and the crew managed to successfully abort a potential disaster.

The second part of the flight from the point of return to the Tokyo (Narita) airport was less dangerous. Engines of modern two-engined automated aircraft are quite powerful, and one normally operating engine is usually capable of providing the thrust needed for the rest of the flight. But to reduce the risk of an emergency in case of failure of the good engine, all two-engine-aircraft operations manuals require immediate landing once an engine failure is noted. In this situation the crew also acted correctly, assuming the responsibility for the overweight landing.

All these facts confirm the suggestion that if pilots expand their professionalism beyond the immediate tasks needed for a normal flight, they can prevent an unforeseen abnormal situation that could threaten crew-aircraft system safety.

Summary

Aircraft automation is a natural phase in aviation development. It has produced a significant impact on flight safety and crew activity. Safety of modern automated aircraft is very high. Rare accidents of these aircraft are caused mainly by incorrect crew actions. Safe flight operations of automated aircraft require that pilots have the basic skills and professional qualities and the ability to operate computerized aircraft automation systems.

To complete a flight successfully, an automated aircraft pilot must have a professional knowledge of the airplane's automatic systems and their functions, control, indications, and operations. Aircraft automation is logically designed, simple to understand, and safe if operated properly.

Continuous monitoring of the flight path and aircraft system parameters is one of the most important flight safety conditions.

Automated aircraft have electronic warning systems that inform pilots about possible threats to flight safety. Immediate and correct pilot reaction to all warning indications can save people's lives as well as the aircraft.

Pilots should use the aircraft automation as their assistant. The automation is a servant, not a master. Servants never command. All decisions in flight should be made by pilots and should be based on all available information. Timely and correct decision making is the main task of the flight crew.

Crew communication and coordination are critical for safe automated aircraft flight. Both pilots must understand and approve all actions performed in the cockpit. The cabin crew is an important part of the entire aircraft staff. Cooperation and coordination between pilots and flight attendants in flight are important factors in flight safety.

Certain physiological and psychological factors may reduce the pilot's abilities to reliably perform flight-related tasks. The pilot should be aware of these factors and be able to eliminate, reduce, or compensate for their effects during flight.

Crew training for automated aircraft is a complex task. Computerized devices are used for ground and flight training. In addition to technological and flight skill training, crew resource management training is required for automated aircraft crews. Aircraft automation requires also some additional flight crew training. To be effective, all types of crew training require well-prepared trainees, optimal programs, and good instructors.

Progress in the aviation industry is developed continuously, as is most other technology. Timely improvements in crew training, crew-aircraft interface, and aircraft design can be fruitful in automated flight efficiency and safety.

Automated aircraft disasters can be prevented. Crew skill and professionalism remain the main resources able to ensure the safe completion of every flight in normal and abnormal conditions. Although aviation is an industry that involves adherence to a lot of rules, instructions, and documentation, it is impossible to develop instructions for all cases of life. The pilot's intellect is the main tool to be used for making good decisions and realizing them during flight.

Bibliography

Airbus Industries, *A310 Flight Crew Operating Manual*, Feb. 1995.

Bainbridge, L., Department of Psychology, University College, London, "Ironies of automation," in *IFAC Analysis, Design and Evaluation of Man-Machine Systems*, Baden-Baden, Germany, 1982, pp. 129–135.

Beaty, D., "The worst six accidents over the last 20 years," *Inter Pilot* 23–29 (June 1992).

Beaudan, E., "Modern flight deck," *Canadian Aviation* 45–46 (Nov. 1989).

The Boeing Company, *Boeing 777 Operations Manual*, Dec. 10, 1997.

The Boeing Company, *Boeing 767 Operations Manual*, Oct. 8, 1999.

Braune, R., and D. M. Fadden, "Flight deck automation today—where do we go from here?" *SAE Technical Papers*, Series 871823, 1987, pp. 141–149.

Bresley, B., and J. Egilsrud, "Enhanced ground proximity warning system," *Airliner* (Boeing Commercial Airplane Group) 1–13 (July–Sept. 1997).

Chittick, J., "Preferential scheduling for aircrew can help address problem of short-term accumulated fatigue," *ICAO Journal* 16–17 (April 1998).

"Conclusions from report on CFIT accident near Kathmandu," *ICAO Journal* **48**(7):23–26 (1993).

Condom, P., "Airborne collision-avoidance systems solutions in sight," *Interavia* **41**(2):197–198 (1986).

Eduoard, P., D. Ivanoff, and J. Dujon, "FFCC—the long evolution," *ICAO Bulletin* 20–22 (July 1981).

Edwards, E. "Human factors in aviation. Part 3: Some contemporary issues," *Aerospace* 14–19 (Oct. 1985).

Farrell, E. P., "Interacting with new technology in the modern flight deck—the airline pilots' view," *SAE Technical Papers*, Series 872391, 1987, pp. 237–241.

Funk, K., B. Lyall, and V. Riley, *Perceived Human Factors Problems of Flightdeck Automation: A Comparative Analysis of Flightdeck with Varying Levels of Automation,* FAA Grant 93-G-039, 1996, Phase 1; available on the Internet at http://flightdeck.ie.orst.edu/FDAI/Phase1/phase1.html.

Helmreich, R. L., "CRM training primary line of defense against threats to flight safety, including human error," *ICAO Journal* 6–10 (June 1999).

Hornby, A. S., *Oxford Advanced Learner's Dictionary of Current English,* 5th ed., Oxford University Press, Oxford, England, 1995.

Hughes, D., "Glass cockpit study reveals human factors problems," *Aviation Week & Space Technology* 32–34 (Aug. 7, 1989).

"Human factors research aids glass cockpit design effort," *Aviation Week & Space Technology* 34–36 (Aug. 7, 1989).

International Transportaion Safety Association (ITSA), *Cockpit Automation Management*, 2000; available on the Internet at http://www.itsasafety.org/ITSA/cockpit.htm.

Julian, K., "Preventing midair collisions," *High Technology* **5:**48–53 (July 1985).

Kaiser, J., FO DFW, Chairman, Accident Investigation Committee, *Flight 965: Accident Investigation Summary;* available on the Internet at http://alliedpilots.org/pub/flightline/nov-1996/flt-965.html.

Kern, T., *Redefining Airmanship*, McGraw-Hill, New York, 1997.

Kozlov, V. V., O. A. Kosolapov, V. I. Zorile, and I. I. Medintsev, "Human factors: Psychophysiologically dangerous factors of flight, and their prevention" (in Russian), Aviation Accident Investigation Society, Moscow (2000).

Krause, S. S., *Aircraft Safety: Accident Investigations, Analyses & Applications,* McGraw-Hill, New York, 1996.

Ladkin, P. B., *AA965 Cali Accident Report near Buga, Colombia, Dec. 20, 1995,* prepared Nov. 6, 1996, for the World Wide Web, Universität Bielefeld, Germany; available on the Internet at http://www.rvs.uni-bielefeld.de/publications/Incidents/DOCS/FBW.html#Cali.

Ladkin, P. B., *Extracts from UK AAIB Report 4/90 on the 8 January 1989 Accident of a British Midland B737-400 at Kegworth, Leicestershire, England*; available on the Internet at http://www.open.gov.uk/aaib/gobme/gobmerep.htm.

Manningham, D., "The cockpit: A brief history," *Business & Commercial Aviation* 56–59 (June 1997).

McMillan, C., and C. Peterson, "Predictive windshear system," *Airliner* (Boeing Commercial Airplane Group) 14–23 (July–Sept. 1997).

Moore, P., and D. M. Page, "Worldwide navigation into the 21st century—an airline view," *The Journal of Navigation* **40**(2):158–163 (1987).

NASA's Ames Research Center at Moffet Field, "Sleep strategy reduces fatigue risk," *Flight International* **129**(4003):9 (1986).

O'Hare, D., and S. N. Roscoe, *Flightdeck Performance: The Human Factor*, Iowa State University Press, Ames, 1990.

Orasanu, J., "Lessons from research on expert decision making on the flight deck," *ICAO Journal* **48**(7):20–22 (Sept. 1993).

Patterson, R. A., *Flight Control Technology for Current/Future Transport Aircraft*, AIAA-84-2941, American Institute of Aeronautics and Astronautics, 1984, pp. 1–4.

Ponomarenko, V. A., and N. D. Zavalova, *Aviation Psychology* (in Russian), Moscow Institute of Aviation and Space Medicine, 1992.

Proctor, P. "Turbulence detector shows promise," *Aviation Week & Space Technology* 73–74 (July 27, 1998).

Risukhin, V. N., *Crew Efficiency Factors Analysis, and Air Transportation Flight Scheduling Method Development* (in Russian), candidate of engineering science dissertation, Academy of Civil Aviation, Leningrad, Russia, 1988.

Roscoe, S. N., *Aviation Psychology*, Iowa State University Press, Ames, 1990.

Shappell, S. A., and D. A. Wiegmann, *The Human Factors Analysis and Classification System—HFACS*, DOT/FAA/AM-00/7, 2000.

Sheehan, J., *The Tyranny of Automation*, 1995; available on the Internet at http://www.faa.gov/AVR/NEWS/Previous/autom.htm.

Small, R. L., et al., "A future direction in pilot training," paper presented at International Aviation Training Symposium (IATS), Sept. 28–30, 1999, Oklahoma City.

Sogame, H., and P. B. Ladkin, Aircraft Accident Investigation Report 96-5, prepared Nov. 17, 1996, for the World Wide Web; available on the Internet at http://www.rvs.uni-bielefeld.de/publications/Incidents/DOCS/ComAndRep/Nagoya/nagoyarep/Nagoya-top.html.

Taylor, R. L., *Instrument Flying*, 4th ed., McGraw-Hill, New York, 1998.

Underwood, D., "Glass—work," *Canadian Aviation* 27–30 (July 1985).

Wiener, E. L., "Cockpit resource management and flight training for the advanced-technology cockpit," *ICAO Journal* 18–19 (Sept. 1993).

Information about aircraft accidents described in this book to illustrate crew errors is available on the Internet:

A300-600 accident in Nagoya (Chap. 1): http://www.rvs.uni-bielefeld.de/publications/Incidents/DOCS/ComAndRep/Nagoya/nagoyarep/Nagoya-top.html

A310 accident in Bucharest (Chap. 2): http://aviation-safety.net/database/1995/950331-0.html

A320 accident in Warsaw (Chap. 4): http:/www.rvs.uni-bielefeld.de/publications/Incidents/DOCS/ComAndRep/Warsaw/Warsaw-report.html

Boeing 757 accident in Cali (Chap. 5): http:/www.rvs.uni-bielefeld.de/publications/Incidents/DOCS/FBW.html#Cali.

A310 accident in Siberia (Chap. 6): http://aviation-safety.net/database/1994/940323-0.htm, and http://planecrashinfo.com/1994.htm

A310 accident in Kathmandu (Chap. 7): http://aviation-safety.net/database/1992/920731-0.htm

Boeing 737 accident in East Midlands (Chap. 8): http:/www.open.gov.UK/aaib/gobme/gobmerep.htm

Index

A

Abnormal checklist, 88
ACAS (see Airborne collision avoidance system)
Accidents, automation-related (see Mishaps, automation-related)
ADF (see Automatic direction finder)
Adverse mental states, 18
Adverse physiological states, 18
Advisories (EICAS), 78
Advisory mode (ECAM), 85
Aerodrome pressure setting, incorrect, 211
Aérospatiale, 73
AFDS (see Autopilot flight director system)
AIMS (airplane information management system), 162
Air traffic controllers, 248–249
Air Transport Association of America (ATA), 142
Airborne collision avoidance system (ACAS), 11, 142–145
Airbus, 44
Airbus A300-600, 11–14, 22–33, 49, 51, 74–75, 79–80
Airbus A310, 11, 47, 49, 51, 74–75, 79–80, 106
Airbus A320, 49
Airbus A330, 49
Airbus A340, 49
Aircraft automation, 5
 and aviator skills, 33–34
 flight-operation, problems related to, 10–14
 history of, 5–7
 opportunities created by, 7–10
 unsuccessful attempts to override, 28–29
Aircraft bank attitude, 289–290
Aircraft configuration, improper, 211–212
Aircraft stability, compromised, 208–209
Airplane information management system (AIMS), 162
Airspace utilization, enhancement of, 293–294
Airspeed indications (PFD), 103–106
Airspeed vector, 104–105
Alpha floor protection, 23, 47
Altitude capture mode (AFDS), 167
Altitude hold mode (AFDS), 168
Altitude indications (PFD), 107–109
Altitude-based alerts (GPWS), 135–136

Index

Analog signals, 41
Angle of attack (AOA), 25, 47, 289
Approach errors, 210–221
  aerodrome pressure setting, incorrect, 211
  air disaster involving, 214–221
  aircraft configuration, improper, 211–212
  communication problems, 212–214
  crew coordination, unsatisfactory, 212
  database information, incorrect, 210–211
Approach mode, 113, 172
Approach programming (of FMS), 202–203
APU (see Auxiliary power unit)
ASRS (see Aviation Safety Reporting System)
ATA (Air Transport Association of America), 142
Attention failures, 17
Automatic direction finder (ADF), 190–191
Automatic (term), 6
Automatic trim, 40
Automation control interface, 173–178
Automation degradation, 233–235
Automation (term), 6
Autopilot (automatic pilot), 5–6
Autopilot flight director system (AFDS), 42–45, 109, 163–174, 229
  common modes of, 171–173
  lateral modes of, 169–171
  longitudinal modes of, 166–169
Autothrottle system, 8, 44–46
  and AFDS, 163
  control of, 176–177
  failure of, 56–57, 90–92
  operational modes of, 173
Auxiliary power unit (APU), 48, 54
Aviation Safety Reporting System (ASRS), 8, 266
Awareness, electronic devices designed to increase flight crew, 72–74

B

Bank angle, 108–109
Barometric altitude indications, 107
Big-team collaboration (see Teamwork)
Boeing, 43, 44
Boeing 737, 49
Boeing 747-400, 49
Boeing 757, 8, 11, 49, 74, 146
Boeing 757/767, 75, 76, 86
Boeing 767, 11, 43, 49, 74
Boeing 777, 10, 46, 47, 49, 86–88, 102–103, 106, 141, 161–162, 232
Bucharest (Romania) A310 disaster (1995), 55–62

C

Cabin decompression, 244
Cali (Colombia) Boeing 757 disaster (1995), 146–154
Cathode-ray-tube (CRT) displays, 51, 70–75
Cautions (EICAS), 77
CDUs (see Control display units)
Checklists, electronic, 87–89
Closed loop pilot training, 282
Cockpit (see Flight deck)
Cockpit voice recorder (CVR), 23
Common modes (AFDS), 171–173
Communication, 6, 212–214
Communication pages (MFD), 87
Complacency, 18, 276
Computer(s), 5, 9–10
  and cockpit design, 70–71
  flight warning, 80
  flight-control, 43
  primary flight, 41, 42
  thrust control/management, 44
Control:
  flight path (see Flight path control)
  maintaining continuous, 62–63
Control display units (CDUs), 75, 175–176, 194–200
Control wheel steering (CWS), 164

Index

Controls, automated aircraft, 37–42
  analog vs. digital signals in, 40–41
  development of, 38–40
  fly-by-wire control system, 38, 41–42
  primary vs. secondary flight controls, 39
Crew:
  collaboration among, 246–248
  electronic devices designed to increase awareness of, 72–74
  and maintenance personnel, 249–251
  and task allocation, 236–245
  training of, 276–283
  (*See also* Pilot[s])
Crew coordination:
  principle of, 229–230
  unsatisfactory, 212
Crew management, 284–287
Crew pairing, 284
Crew proficiency, 97–98
Crew resource management training, 279–280
Crew resource mismanagement, 19, 59–60, 181–182
CRT displays (*see* Cathode-ray-tube displays)
Culture, 21
CVR (cockpit voice recorder), 23
CWS (control wheel steering), 164

D

Dampers, 40
Decision errors, 17, 218–219
Decision-making training, 282–283
Decompression, cabin, 244
Deliberate distortion of motivation, 273–274
Design, automated aircraft, 37–63
  and aircraft motion control errors, 55–62
  automatic flight-control system, 42–48
  controls, automated aircraft, 37–42
  power plant, automated aircraft, 48–54

Development of aircraft automation:
  controls, 38–40
  flight deck, 70–72
DFDR (*see* Digital flight data recorder)
Dhaka (Bangladesh) A310 mishap (1997), 89–97
Digital flight data recorder (DFDR), 23, 57
Digital signals, 41
Director control, 162
Dispatch deviation guide, 78
Display select panel (DSP), 86–88
Displays, 51–52
Distance measuring equipment (DME), 191–192
Dominant condition, pilot, 272–273
Doors, aircraft, 245–246
Drowsiness, 273
DSP (*see* Display select panel)
Dynamometric rods, 43

E

EADIs (electronic attitude director indicators), 74
East Midlands (England) Boeing 737-400 aircraft disaster (1989), 252–257
ECAM (*see* Electronic centralized aircraft monitor)
Efficiency, 261–263
  and crew-aircraft-environment system, 283–294
  (*See also* Training)
EFIS (*see* Electronic flight instrument system)
EGPWS (*see* Enhanced ground proximity warning system)
EGT (exhaust gas temperature), 53
EHSIs (electronic horizontal situation indicators), 74
EICAS (*see* Engine indication and crew alerting system)
Electronic attitude director indicators (EADIs), 74
Electronic centralized aircraft monitor (ECAM), 75, 79–86
  left display of, 81–83
  operation modes of, 84–85
  right display of, 83–84

Electronic checklists, 87–89
Electronic flight instrument system (EFIS), 8, 9, 74, 102–103, 107, 117–119
Electronic horizontal situation indicators (EHSIs), 74
Engine indication and crew alerting system (EICAS), 74, 76–79, 85–89
  abnormal-situation messages in, 77–78
  lower display of, 78–79
  upper display of, 77
Engine pressure ratio (EPR), 52
Engines, 6, 8, 48–54
Enhanced ground proximity warning system (EGPWS), 11, 139–140
EPR (engine pressure ratio), 52
Ergonomic factors, 263–264
Errors, 16–17
  in aircraft motion control, 55–62
  in aircraft status monitoring, 89–97
  crew coordination, 251–257
  in flight path control, 178–183
  in navigation, 205–221
  skill-based, 59, 150–151, 181, 218–219
  in use of flight parameter indication, 119–129
  in use of warning systems, 145–154
Exceptional violations, 18, 181
Exhaust gas temperature (EGT), 53
Extreme speed surges, protection against, 46

F

FAA (*see* U.S. Federal Aviation Administration)
Failure mode (ECAM), 85
Failure to correct a known problem, 20
Failure to switch for a new activity, 274–275
Fatigue, 266–267
FCCs (flight-control computers), 43
FCOM (flight crew operations manual), 13
FCU (*see* Flight-control unit)
FD (*see* Flight director)
FD bars (*see* Flight director bars)
Final impulse phase, 266
Fire, 243, 244
Flight-control computers (FCCs), 43
Flight-control unit (FCU), 43, 166
Flight crew (*see* Crew)
Flight crew operations manual (FCOM), 13
Flight deck, 69–98
  aircraft systems status electronic indication device in, 76–85
  case study of aircraft-status monitoring errors in, 89–97
  crew efficiency and design of, 288
  development of, 70–72
  electronic centralized aircraft monitor (ECAM) in, 79–86
  engine indication and crew alerting system (EICAS) in, 76–79, 85–89
  flight crew awareness, electronic devices designed to increase, 72–74
  "glass cockpit" in, 74–76
  and isolation of aircraft system failures, 85–86
  multifunctional display in, 86–89
  task allocation in, 230–236
  technology used in, 69–76
Flight director (FD), 22–24, 44
Flight director (FD) bars, 163, 164
Flight duty, duration of, 267–268
Flight envelope protection, 46–47
Flight level change mode (AFDS), 168
Flight management computer (FMC), 105, 109, 175, 177, 194
Flight management system (FMS), 8, 9, 175, 176, 194–205
Flight mode annunciator (FMA), 23, 25, 43, 45, 111, 177–178
Flight path angle (FPA), 110–111
Flight path control, 161–184
  automatic flight director system (AFDS), 163–174

Index 313

Flight path control (*Cont.*):
  and automation control interface, 173–178
  errors in, 178–183
  manual control, 162
Flight path parameter electronic indications, 101–129
  crew errors resulting from incorrect use of, 119–129
  electronic flight instrument system (EFIS), 102–103, 117–119
  navigation display (ND), 113–117
  primary flight display (PFD), 103–113
Flight path position indications (PFD), 110–111
Flight path vector (FPV), 44, 110–111
Flight simulators, 73
Flight technique failures, 17
Flight warning computers (FWCs), 80
Fly-by-wire control system, 38, 41–42, 162
FMA (*see* Flight mode annunciator)
FMC (*see* Flight management computer)
FMS (*see* Flight management system)
FPA (*see* Flight path angle)
FPV (*see* Flight path vector)
Fuel flow, 53
FWCs (flight warning computers), 80

G

"Glass cockpit," 70, 74–76
Global positioning system (GPS), 170, 189
Go levers, 12, 22–23, 29, 32, 43, 45
Go-around mode, 12, 172–173
GPS (*see* Global positioning system)
Ground proximity warning system (GPWS), 7, 111, 134–142
  look-ahead terrain alerts in, 136–137

Ground proximity warning system (*Cont.*):
  radio-altitude-based alerts in, 135–136
  terrain avoidance maneuver in, 138–140
  terrain display in, 137

H

Heading hold mode (AFDS), 169
Heading selection mode (AFDS), 169–170
HFACS (human-factors analysis and classification system), 14
History of aircraft automation (*see* Development of aircraft automation)
Hold mode (autothrottle system), 173
Horizontal navigation, 195–196, 198–199
Horizontal situation indicator (HSI), 75
Human failure, levels of, 15
Human role in automated aircraft flight (*see* Task allocation; Teamwork)
Human-factors analysis and classification system (HFACS), 14
Human-factors analysis of aviation mishaps, 14–33
Hydraulic actuators, 43

I

Idle mode (autothrottle system), 173
Illusion of located target, 272
ILS (*see* Instrument landing system)
Industrial revolution, 5
Inertial reference system (IRS), 188–189, 201
  update, improper, 209–210
Instrument landing system (ILS), 23, 94, 111, 112, 192–193
IRS (*see* Inertial reference system)

J

Jet engines (*see* Engines)
Jet lag, 268–269

K

Kathmandu (Nepal) A310 aircraft disaster (1992), 214–221
Kern, Tony, 14, 15, 22
Knowledge, lack of professional, 29–30
Known problem, failure to correct, 20

L

Landing reference speed, 105
Lateral navigation mode (AFDS), 170
LCDs (see Liquid crystal displays)
Le Bourget Exhibition, 73
Liquid crystal displays (LCDs), 52, 70, 74
Localizer, 171
Localizer beam, 171
Localizer mode (AFDS), 171
Look-ahead terrain alerts (GPWS), 136–137

M

Mach number, 106
Maintenance personnel, 249–251
Manual flight path control, 162
Manual mode (ECAM), 85
Map mode (navigation display), 113–116
Marker transmitters (markers), 193
Maximum maneuvering speed, 105
Maximum speed, 106
MCP (see Mode control panel)
MD-11, 49
MEL (see Minimum equipment list)
Memory failures, 17
Mental states, adverse, 18
MFD (see Multifunctional display)
Microburst, 140
Minimum equipment list (MEL), 78, 250–251
Minimum maneuvering speed, 105
Minimum speed, 105–106
Mishaps, automation-related, 10–33
  Bucharest, Romania (1995), 55–62

Mishaps, automation-related (*Cont.*):
  Cali, Colombia, 146–154
  Dhaka, Bangladesh (1997), 89–97
  East Midlands, England (1989), 252–257
  Grand Canyon (1956), 143
  human-factors analysis of, 14–33
  Kathmandu, Nepal (1992), 214–221
  Nagoya, Japan (1994), 12–14, 22–33
  San Diego (1978), 143
  Siberia (1994), 179–183
  Tokyo, Japan (1998), 295–297
  Warsaw, Poland (1993), 120–129
Mixed-fleet flight operations, 284–285
Mode control panel (MCP), 43, 109, 111, 166–169, 174
Mode selection, forgetful, 207–208
Monotony, 269
Motivation, 18, 181–182, 273–274
Multifunctional display (MFD), 86–89

N

N_1 parameter, 52–53
N_2 parameter, 53
Nagoya (Japan) A300-600 aircraft disaster (1994), 12–14, 22–33
National Aeronautics and Space Administration (NASA), 8, 72
Navigation, aircraft, 187–222
  approach errors, 210–221
  approach phase, 202–205
  control display unit navigation pages, 199–201
  electronic systems, 188–194
  errors in, 205–221
  in flight, 197–201
  flight-management-system controlled navigation, 194–205
  global positioning system, 189
  horizontal navigation, 195–196, 198–199
  inertial reference system, 188–189

Navigation, aircraft (*Cont.*):
 initialization errors, 205–206
 position calculation update, aircraft, 201
 preparation phase, 195–197
 radio systems, 189–194
 route errors, 206–210
 vertical navigation, 196–199
Navigation display (ND), 113–117
 error avoidance with, 116–117
 map mode expanded abilities, 114–116
 operation modes of, 113–114
 parameter indication controlled on, 118–119
 terrain on, 136
 windshear alerts on, 141
New activity, failure to switch for a, 274–275
Normal mode (ECAM), 84–85

O

Operational processes, 21
Operations, 21
Operators:
 substandard conditions of, 18–19
 substandard practices of, 19
Organizational climate, 21
Organizational factors, 20–21, 31, 60–61, 127, 152–153, 182–183, 220, 256
Other traffic display (TCAS), 144–145
Overconfidence, 18
Oversight, 21

P

Panel displays, 9
Perceptual errors, 17
Personal readiness, 19
Personality traits, 18
PFD (*see* Primary flight display)
Physical or mental limitations, 18–19
Physiological states, adverse, 18
Pilot dominant condition, 272–273
Pilot(s), 263–276
 air traffic controllers, collaboration with, 248–249

Pilot(s) (*Cont.*):
 characterizing the condition of, 264–265
 complacency in, 276
 deliberate distortion of motivation by, 273–274
 dominant condition, pilot, 272–273
 drowsiness in, 273
 and duration of flight duty, 267–268
 error, causes of pilot, 263–264
 and failure to switch for a new activity, 274–275
 fatigue in, 266–267
 flight-relevant factors influencing, 265–270
 and illusion of located target, 272
 incapacitation of, 58
 informational models used by, 101
 jet lag in, 268–269
 and monotony, 269
 physiological factors influencing, 270–276
 psychological direction of, 275–276
 relief, 266–267
 sleep time for, 268
 and stress, 269–270
 tasks of, 238–239
 training of, 276–283
 understanding of automatic flight-control functions by, 47–48
 (*See also* Priority lists, pilot's)
Pitch angle, 108
Pitch attitude, 109
Pitch mode, 30
Plan mode (navigation display), 114
Planned inappropriate operations, 19–20
Policies, 21
Poor-choice errors, 17
Position bars, 193
Power plant, automated aircraft, 48–54
 auxiliary power unit, 54
 propulsion system, 48–54

316 Index

Preconditions for unsafe acts, 18
Predictive windshear system
(PWS), 140–141
Preset function mode (AFDS), 168
Primary flight computer, 41, 42
Primary flight controls, 39
Primary flight display (PFD), 23,
43, 75, 103–113
  airspeed indications on, 103–106
  altitude indications on, 107–109
  bank angle on, 108–109
  flight path position indications
    on, 110–111
  miscellaneous indications on,
    111–113
  parameter indication controlled
    on, 117–118
  steering indications on, 109–110
  time-critical warnings on, 111
Priority lists, pilot's:
  aircraft position, continuous and
    complex verification of,
    221–222
  continuous control, remaining
    in, 62–63
  flight path monitoring and
    control, 183–184
  flight-relevant information,
    procurement/analysis of, 129
  professionalism, expanded,
    297–298
  system status considerations,
    aircraft, 97–98
  teamwork, 257–258
  warning signals, immediate and
    correct reaction to, 154–155
Problem-solving errors, 17
Procedural errors, 17
Procedures, 21
Profitability, 5
Propulsion system, 48–54
  parameters, engine, 51–53
  thrust controls, engine, 51
  thrust reverser, engine, 53–54
Proximate traffic display (TCAS),
144–145
Psychological factors, 270–276
Purser, 241–242
PWS (*see* Predictive windshear
system)

Q

Quick reference handbook
(QRH), 86, 88

R

RA (resolution advisory), 144
Radar, weather, 194
Radio altitude indications,
107–108
Radio navigation systems,
190–194
  automatic direction finder,
    190–191
  distance measuring equipment,
    191–192
  frequency range for, 191
  instrument landing system,
    192–193
  marker signal receiver, 193
  transponder, 193–194
  weather radar, 194
Radio-altitude-based alerts
(GPWS), 135–136
Reactive windshear system (RWS),
140
Reason, James, 15, 16
Recurrent training, 278–279
Relief pilots, 266–267
Resolution advisory (RA), 144
Resource management decisions,
20–21
Roll envelope bank angle
protection, 47
Route errors, 206–210
  aircraft stability, compromised,
    208–209
  inertial reference update,
    improper, 209–210
  mode selection, forgetful,
    207–208
  waypoint estimate, incorrect, 207
  weather avoidance, belated, 209
Routine violations, 18
  of aircraft operation require-
    ments, 59
  of flight operation procedures,
    58
RWS (*see* Reactive windshear
system)

Index 317

S

Safety tasks, 242–245
SDAC (system data analog/digital converter), 80
Secondary flight controls, 39
Semiautomatic flight path control, 162
Service tasks, 245
SGUs (symbol generator units), 80
Shappell, Scott A., 14, 16
Siberia A310 disaster (1994), 179–183
SID (standard instrument departure), 206
Skill-based errors, 17, 29, 59, 150–151, 181, 218–219
Skills, pilot, 33–34
Sleep, 268
Smoke, 243
Speed, 103–106
Speed mode (autothrottle system), 173
Speed surges, protection against extreme, 46
Speed-trend vectors, 9
Sperry Company, 74
Stall protection, 46
Standard instrument departure (SID), 206
Standard terminal arrival route (STAR), 202
Status messages (EICAS), 78
Steering indications (PFD), 109–110
Stick shaker, 46, 91, 106
Stress, 269–270
Structure, organizational, 21
Substandard conditions of operators, 18–19
Substandard practices of operators, 19
Supervision, unsafe, 19–20
Supervisory violations, 20
Symbol generator units (SGUs), 80
System data analog/digital converter (SDAC), 80
System failures, isolation of, 85–86

T

TA (traffic advisory), 144
Takeoff/go-around (TO/GA) switches, 43
  (*See also* Go levers)
Takeoff mode (AFDS), 171–172
Task allocation, 228–245
  and automation degradation, 233–235
  cabin crew, 241–245
  cockpit crew, 236–241
  concept of, 228–229
  and crew coordination principle, 229–230
  in flight deck, 230–236
  and limitations of automation utilization, 231–233
TCAS (*see* Traffic alert and collision avoidance system)
Teamwork, 245–258
  between cockpit crew and cabin crew, 246–248
  errors due to lack of, 251–257
  between flight crew and maintenance personnel, 249–251
  between pilot and air traffic controllers, 248–249
Terrain avoidance maneuver, 138–140
Terrain display (GPWS), 137
Thrust control computer, 44
Thrust levers, 12, 51
Thrust management computer, 44
Thrust mode (autothrottle sytem), 173
Thrust reference mode (autothrottle system), 173
Thrust reverser, engine, 51, 53–54
THS (*see* Trim-of-horizontal-stabilizer)
Time-critical warnings (PFD), 111
TO/GA switches (*see* Takeoff/go-around switches)
Tokyo (Japan) A310 aircraft incident (1998), 295–297
Traffic advisory (TA), 144
Traffic alert and collision avoidance system (TCAS), 109–110, 143–145

Training, 276–283
  closed loop pilot, 282
  of crew, 276–283
  crew resource management, 60, 279–280
  decision-making, 282–283
  inadequate, 126–127, 151, 182, 219–220, 255–256
  lack of, 30–31
  philosophy of, 280–282
  recurrent, 278–279
  transition flight, 277–278
Training programs, 15–16
Transducer, 41–42
Transition flight training, 277–278
Transponder, 193–194
Trim-of-horizontal-stabilizer (THS), 23–26
Turbulence, 244–245

U

Unsafe supervision, 19–20
U.S. Federal Aviation Administration (FAA), 8, 75, 140

V

Vertical navigation, 196–199
Vertical speed, 106
Vertical speed mode (AFDS), 166–167

Very-high-frequency (VHF) signals, 191
Violations, 16–18
VNAV mode (AFDS), 168
VOR mode, 113–114, 170

W

Warning systems, electronic, 133–155
  airborne collision avoidance system, 142–145
  crew errors in use of, 145–154
  ground proximity warning system (GPWS), 134–142
Warnings (EICAS), 77
Warsaw (Poland) A320 aircraft accident (1993), 120–129
Waypoint entry, incorrect, 205
Waypoint estimate, incorrect, 207
Waypoint omission, 206
Weather avoidance, belated, 209
Weather minima conditions, 16
Weather radar, 194
Weight value entry, incorrect, 206
Wiegmann, Douglas A., 14, 16
Windshear, 47, 140
Windshear alerts (GPWS), 140–142
Windshear Training Aid, 140

About the Author

Vladimir Risukhin is Russia's leading authority on human factors and automation in aviation. His specialties include, but are not limited to, training pilots to transition from the larger aircrews of traditional aircraft to the two-person cockpits of Western planes such as the B767, the B777, and the Airbus. He has 25 years of experience in aviation instruction around the world, and has conducted dozens of automated aircraft crew training and crew resource management (CRM) courses with a focus on automation integration.

www.ingramcontent.com/pod-product-compliance
Lightning Source LLC
Chambersburg PA
CBHW011953150426
43198CB00019B/2921